Albert Bledsoe Herrick

Modern Switchboards and the Appliances used Thereon

Albert Bledsoe Herrick

Modern Switchboards and the Appliances used Thereon

ISBN/EAN: 9783337337667

Printed in Europe, USA, Canada, Australia, Japan

Cover: Foto ©berggeist007 / pixelio.de

More available books at **www.hansebooks.com**

MODERN SWITCHBOARDS

AND THE APPLIANCES USED THEREON; TOGETHER WITH AN HISTORICAL RÉSUMÉ OF EARLY PRACTICES AND EXPEDIENTS, INDICATING THE ADVANCE RECENTLY MADE IN THIS CLASS OF ELECTRICAL APPARATUS; AND DATA ON APPROVED METHODS OF CONSTRUCTION

BY

ALBERT B. HERRICK

ILLUSTRATED BY NUMEROUS CUTS, DRAWINGS AND DESIGNS, ESPECIALLY PREPARED FOR THIS PUBLICATION

THE CUTTER ELECTRICAL AND MANUFACTURING COMPANY, PHILADELPHIA, U.S.A., MDCCCXCVIII.

COPYRIGHT, 1898
THE CUTTER ELECTRICAL AND
MANUFACTURING COMPANY

ALL RIGHTS RESERVED

PRICE, THREE DOLLARS

PRESS OF
EDWARD STERN & CO., INC.
PHILADELPHIA.

CONTENTS

INTRODUCTORY.

The advance of the art of switchboard construction. Old methods of operating. New methods, ... 7

CHAPTER I.

CIRCUIT BREAKING DEVICES. The early methods of severing the circuit. Various switch forms. Slow break snap switches and switches with auxiliary contacts. Physical properties of contacts. Types of switch contacts; American, English and German types. Mechanical and electrical properties of contact surfaces, ... 12

CHAPTER II.

SWITCHBOARD CONSTRUCTION. Switchboard attendance. Gallery construction and detail. Relation of switchboards to distributing systems. Fireproof construction. Exposure. Framing and insulation of switchboards. Method of connecting conductors and bus bars. Conductor losses. Material for electrical engineering. Impurities of copper. Dimensions of bus bars. Weights and current capacities. Alloys and their conductivity. Brass and other copper alloys. Switchboard material and their necessary properties. Insulators—wood, slate, marble and onyx. Method of drilling. Special method of switchboard construction. Insulating in high tension switchboards, ... 21

CHAPTER III.

SWITCHBOARD APPLIANCES. Potential measurements. Requirements of switchboard instruments. Method of testing for errors. Checking methods. Instrument movements. Solenoid. Permanent magnet type. Instruments actuated by rise in temperature. Astatic voltmeters. Recording instruments. Voltmeters. Comparative pressure indicators. Indicating wattmeters. Integrating wattmeters. Dynamo galvanometers. Dynamo regulators. Automatic regulators, ... 41

CHAPTER IV.

PROTECTIVE DEVICES. Fuses. Physical conditions under which fuses operate. Condition where fuses are useful. Cutter circuit breakers—operation and reliability. Types of I-T-E circuit breakers. Lightning arresters. Characteristics of lightning discharges. Proper method of lightning protection. Magnetic type of lightning arrester. Types of lightning arresters. Non-arcing lightning arresters, ... 60

INTRODUCTORY

THE evolution of the switchboard has necessarily followed the progress of the various systems of distribution, as a necessary adjunct for the controlling and connecting of the different circuits, and it is to-day necessary for the collection, distribution and control of output in any electric light, power or railway system.

In order to fully comprehend the present switchboard practice, an historical *résumé* may be necessary. The earliest attempt at a collection of apparatus for the purpose of controlling a distributing system was at Menlo Park, in 1879, when Edison made the first commercial application of low-tension currents to a multiple arc distributing system. The effect of assembly here was especially the following the diagram of connections with bare copper wires connecting the different plug devices.

The points of serviceability and utility came later on in the art, and the necessity of measuring devices for the current flow and potential, as well as the protecting devices against abnormal flow of current, soon asserted itself.

The first central station erected had the apparatus for circuit and dynamos strewn around the four walls of the station, regardless of utility.

In 1883, Mr. Luther Stierenger, at the Louisville Exposition, designed the distribution circuits so that they and their dependent apparatus were concentrated at one point. All switches, fuses, and bus bars were brought together, and the points of utility and convenience were realized; this afterward developed into switchboard construction.

The placement of the switchboard in the older practice was left as one of the last points to be considered in laying out a station, but now the proper position has been found to have such direct bearing on the facility of handling the apparatus which it controls that the proper placement has become a primary matter, and the engineer has need of all his ability and judgment in considering the conditions which determine the best possible position. Originally wires secured to the wainscoting and plugs and crude instruments were connected in by splices. A fire hazard developed from this combination; the switchboard was then placed at a distance from the wall, and on a skeleton frame, in the construction of which as little wood as possible was used. This sufficiently reduced the hazard until fire-proof methods of construction were developed. At first the bus bars and their connecting cables were kept on the front of the board until back connections were necessitated on the score of safety, space and appearance.

Wood was universally used for supporting instrument cases, regulators and equalizers, as well as for insulation, up to 1889, when, on account of the high fire hazard placed on such construction, slate and porcelain were substituted for wood, as the insulating and supporting structure.

The operation of the early switchboards was comparatively simple, and the attendance and space necessary were not considered items to be taken into account; but on enlarging the field of current distribution systems, careful designing had to be done, in order to make the switchboard as compact, simple and easily operated as possible; also, as the current required by the expanding system

INTRODUCTORY

increased, new methods had to be introduced in handling the dynamos and feeders of increased capacity.

Fig. B shows the dynamo board of the old Adams Street Station, Chicago, Ill., which was an advanced type at the time of its construction, with exposed bus bars and all apparatus assembled on a wooden facing, using wood for insulation.

Fig. C shows the feeder board for the same station, including wooden frame equalizers, which are now relegated to the scrap heap in nearly every station, as external systems of distribution have become more complex and require economical methods to obtain uniform potential on mains, to meet the varying demands on the system.

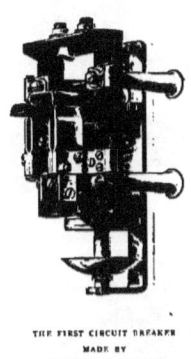

THE FIRST CIRCUIT BREAKER
MADE BY
THE CUTTER COMPANY

Fig. B. DYNAMO BOARD
ADAMS STREET STATION, CHICAGO, ILL.

FIG. C. FEEDER BOARD
ADAMS STREET STATION, CHICAGO, ILL.

CIRCUIT BREAKING DEVICES

N the evolution of circuit breaking devices, it can be clearly shown that it is by the slow process of the survival of the best forms, that the present stage of the art has been reached.

The initial letter illustrates the primitive form of circuit rupturing device, when the conductor itself is severed.

The switching devices that were adopted in the early days were mere enlargements of those used in the feeble carrying current art.

The well-known plug came directly from the telegraph plug, and the first switching devices were derived from the telegraph switch, only enlarged to accommodate the greater volume of current flow.

The necessity for a mechanical device, which could open the circuit without requiring the conductors to be moved, was met by having two circuit terminals end in two insulated plates, the adjacent ends of which were connected together by the insertion of a plug, as shown in Fig. 1. The adoption of this form of device to the heavier current art required larger terminals and more area of plug connection.

FIG. 1

Fig. 2 shows the form commonly used. The functional weakness of this connection was that the adjacent breaking surfaces were too near each other, so that on withdrawing the plug the arc followed, and could be maintained between the terminals.

CIRCUIT BREAKING DEVICES

The arc was extinguished originally by being blown out; but as the current density ran up, other means were used, and it was a familiar sight to see a sand-box handy, so that if the arc was too fierce to be blown out, it could be extinguished by throwing a handful of sand on it. Two plugs were also connected in series, one to break the main circuit, and the other to extinguish the resultant arc.

Fig. 2

The first invention in switching mechanisms was the introduction of a plug and flash plate which were normally depressed below the surface of the plug switch base; but on withdrawing the plug this false insulation plug and flash plate severed the arc by springing between the terminals. Fig. 3 shows the next step where, by separating the circuit terminals, the arc cannot hold; in this switch is also shown the elements of the knife switch, which soon followed.

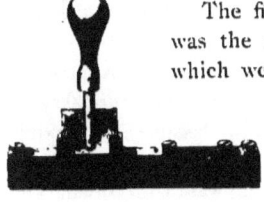

Fig. 3

Fig. 4

The circuit breaking devices were, during the same period, following along other lines which were in their inception a modification of Fig. 4, the strap key, and Fig. 5, the telegraph key. Fig. 6 shows the earliest form adopted from this origin, for heavy current-carrying devices, and was first made in 1879, for headboard switches for the Edison "Z" dynamos.

Fig. 5

CIRCUIT BREAKING DEVICES

In this switch the contact faces consist of two abutting surfaces,  one being rigid on the switch base and the other attached to a lever which forms one terminal of the circuit, being held positively in open or shut position by means of a spring compressing a stop in a two-way jockey plate, as shown.

Fig. 7 shows another basic form, from which was probably developed the design of the switch  shown in Fig. 8, but this development took place at a later period.

FIG. 6

FIG. 7

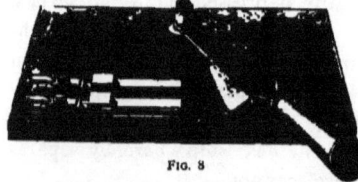

FIG. 8

From the switch shown in Fig. 3 the current-carrying contact became a flexible plate, bearing against a lever which was withdrawn, and here the advance was made of having a wiping contact, and the arc was not drawn on the current-carrying surfaces. Fig. 8 shows the first mechanical form of this switch, and Fig. 9 another form where the pivot forms one terminal. These switches were next designed in the vertical plane, in order to reduce the room occupied, and took the forms shown in Figs. 10 and 11. At this time the high potential large current art required another function in switching

FIG. 9

CIRCUIT BREAKING DEVICES

devices, which was the severing of the current with great rapidity, in order that the arc should not follow after the severing member of the switch.

Fig. 12 shows the first attempt to attain this object. The device consists of two flexible contacts which are connected by a conducting blade, the lower edge of which has an insulated blade which, on the withdrawal of the conducting blade, introduces an insulating medium between the terminals of the switch. Also for rapid removal, on withdrawing the blade, a spring is put under compression, which snaps the connecting blade away from the circuit terminals.

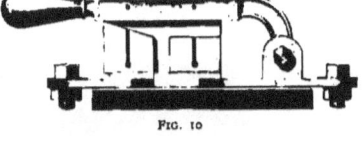

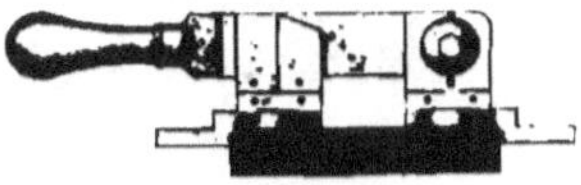

Fig. 10

Fig. 11

The next form of switch, Fig. 13, is an adaptation made from Fig. 8, with the addition of a mechanical device to throw the switch quickly by means of compressing a spring when the handle is thrown over, which will withdraw the blade from its contacts and snap it in open position.

Fig. 12

Fig. 13

After these, numerous devices were adopted, having for their function the putting under tension the blade as it was withdrawn from the terminals or clips. Fig. 14 shows the first form, where the blade is first under tension from a spring on the handle, which

CIRCUIT BREAKING DEVICES

then engages with the blade, withdraws and snaps on being relieved from the contact friction. Fig. 15 is another form where the blade is withdrawn under tension by means of a spiral spring.

Fig. 16 is still another form where

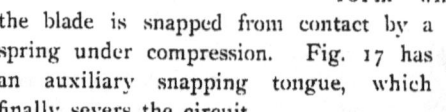
FIG. 14

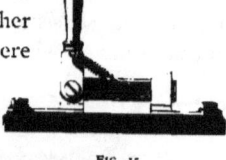
FIG. 15

the blade is snapped from contact by a spring under compression. Fig. 17 has an auxiliary snapping tongue, which finally severs the circuit.

Fig. 18 shows a snap switch where the blade is divided into two parts entering one clip, the upper half leaving the contacts and placing the lower half under tension, which is withdrawn with a snap.

FIG. 16

FIG. 17

Fig. 19 is a variation from Fig. 18, in having the snapping and moving blades enter two different clips, while a tension is placed between them on withdrawing the blade fixed to the handle.

FIG. 18

Fig. 20 shows auxiliary contacts composed of carbons. The purpose of these is to effect the

CIRCUIT BREAKING DEVICES

same result as that practically obtained by quickly snapping the blade from its contact. This arrangement introduces a high resistance in the circuit before severing it, and the points of arcing at breaking the circuit are between the carbons. The result is an arc of high resistance, which is easily extinguished. A fuse has been bridged by a switch blade which will blow after opening

FIG. 19

the switch, and in this way protect the switch terminals. On high inductive circuits with alternating currents, reactive devices have been placed between the terminals of a switch which opens the main circuit, and then the reactive device is cut out. All these arrangements have for their object the reduction in volume of the main current before finally opening the circuit.

FIG. 20

SWITCH TERMINALS

The positions of the parts and members of the switch are largely a matter of convenience, but the method of designing in order to decrease the drop through the switch with the least possible amount of material is one of contacts and their physical properties. This matter has been given a great deal of thought, and it can be said that there is yet no standard form of contact.

The Germans, English and Americans have

FIG. 21

each evolved their own type, each with its distinctive merits. Taking up the original type of contact, which consists of a rigid

17

SWITCH TERMINALS

member sliding between two surfaces, the first form brought out was a blade fitted between two rigid walls of metal, as shown in Fig. 21. This was used in electroplating practices, where the switch was not deteriorated by arcing effects. The next step, Fig. 22, was to have one side of the terminal, into which the blade entered, flexible, and holding the blade against a rigid support in order to allow a slight movement of the contact, due to inequalities of the contact surfaces. Fig. 23 shows where both of these clip surfaces have been made flexible in order to further increase the contact surfaces between the terminal and its blade.

Fig. 22

Fig. 23

Fig. 24 shows another method of securing the flexible contacts to the terminal plate. Fig. 25 shows the same general form of construction, but more flexible in its bearing on the contact surfaces, and the flexible contacts soldered into grooves cut in the terminal plate. Fig. 26 is a form used in small-capacity switches, where the flexibility of contact is obtained by a continuous metal plate bent to form both contact clips. Fig. 27 shows the same result obtained by reversing the conditions found in Fig. 25, the moving arm, in this case, being the flexible contact member and the terminal being the rigid contact member. The object of this arrangement was to

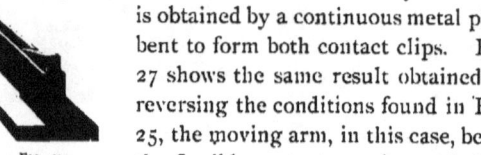

Fig. 25  Fig. 24

Fig. 27 Fig. 26

SWITCH TERMINALS

reduce the drop on the switch, due to the discontinuous connection between the flexible contact member and the terminal. Fig. 28 shows a multiplication of the same general construction as Fig. 25, in order to increase the area of the contact without introducing the inflexibility of one large moving switch blade.

FIG. 28

FIG. 29

Fig. 29 is the element from which the German and English types of switch emanated. They consist of a flexible contact surface, bearing on a rigid terminal surface; in this case the contact is a flexible brush, wiping over the terminal. This has been reduced in heavier sizes of switches to a number of flexible fingers, bearing against the contact terminals, as shown in Fig. 30.

FIG. 30

FIG. 31

Fig. 31 shows the same element of construction, but where the blade is introduced between the rigid terminals. Fig. 32 shows the construction which effects the same flexibility between the different fingers of two contact surfaces, the two surfaces being in this case the separate terminals of the circuit.

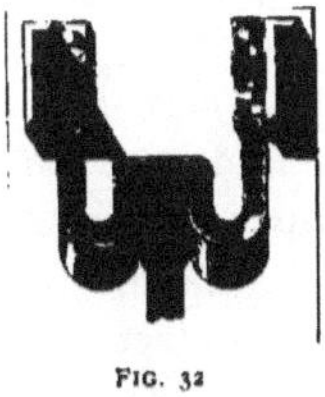

Fig. 33 shows the recent American improvement in contact surfaces where the movable contact member consists of a laminated surface of

FIG. 32

FIG. 33

a great number of individual spring contacts, which, on their intro-

19

duction between contact surfaces, offer individual flexible contact points for carrying the current from the movable member to the terminals, and in this way presenting a large surface under compression.

CONTACT SURFACES

The materials which form the contacts should possess the quality of not mutually abrading each other when rubbing together. Two similar metals are liable to bite and tear the contact surfaces, and they should be selected so as to have different physical characteristics, in order that they will wear well together.

The conductivity of a contact surface is dependent upon three values: the pressure which they bear on each other, the character of surface exposed to the conductivity of current, and the character of metals forming the contact areas.

As a conductor, a contact surface behaves as if it were composed of a multitude of points on a flexible warped surface, and as the pressure is increased, more of the points come in contact, and the pressure plays a more important part in the conductivity of a contact than the area. With a rise in temperature of the terminals, the actual drop between the contact will fall, if there is no local potential due to the metals in the contact setting up a local thermo-electro-motive force, or Peltier effects.

In the case of copper and zinc, and their alloys, there is no appreciable thermo effect, but for lead-aluminium, lead-iron, tin-aluminium, tin-iron, bismuth-iron, bismuth-aluminium, in combinations, there is a very appreciable resistance over normal, due to local counter electro-motive forces; the resistance of contact rises rapidly with time, and all local currents tend to depreciate the surfaces of contact through which they act.

SWITCHBOARD CONSTRUCTION.

LOCATING the switchboard is determined by the relative positions of the operating machinery in the plant. It is desirable that all points of control, namely, the throttle of the engine, the switchboard and the dynamo commutators, be near together, and so arranged that they can be easily reached by an attendant without squeezing past fly-wheels or belts.

In short, do not arrange your apparatus so as to put the attendant under any physical hazard.

When trouble arises, it is necessary to take care of several things at once, and their close proximity to each other renders the attendant more efficient. This is especially true in isolated plants. In moderate-size stations the functions of each attendant are less involved, and the switchboard and generators are generally under one man's care. In the larger stations the general practice is to have one attendant whose only duty is to take care of the switchboard. The modern tendency, both abroad and at home, is to place the switchboard on a gallery or in an elevated position, where the attendant has a full view of the operation of the

FIG. 35

GALLERY DESIGN AND LOCATION

generators under his charge, and can act promptly in a case of emergency.

The following shows a number of designs of gallery constructions; also stairways leading to gallery; both spiral and straight. Designs of railings are shown in Plate 34; Fig. 35 shows a design of a gallery of cantilever construction; Fig. 36 shows a double-decked switchboard, reached by a spiral stairway, and Fig. 37 shows a plain railing for a slightly elevated gallery. Each case of gallery construction is entirely dependent upon the architectural arrangement of the station, and these illustrations are given only to indicate certain practices. In this location of the switchboard, if the dynamo leads are run above the dynamo to an underground system,

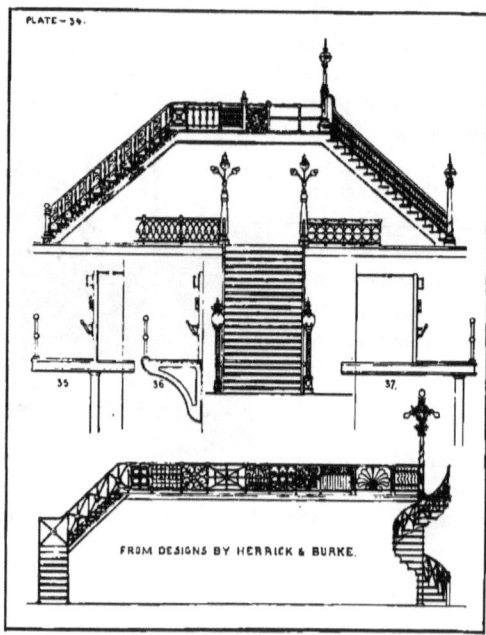

PLATE – 34.

FROM DESIGNS BY HERRICK & BURKE.

or underneath the dynamo to an overhead system of distribution, extra copper in these two cases is not necessary to conduct the current to the gallery. Where the current has to be carried up to the gallery and back again, there is considerable length of conductor used, only on account of the gallery location.

Where the handling of 100,000 amperes is concerned, as in some of the larger three-wire systems, where distribution is underground, the indications are that future practice will be to locate the bus bar in the same plane as the distribution and supply systems, and to operate the switching mechanism from the gallery by mechanical or transmission methods, thereby saving a large expense in conductors and waste of energy in transmission, there being located on the gallery only the measuring instruments and regulating devices, and the levers to operate the circuit-changing switches. Compressed air is used at present for mechanically opening circuits at high speeds at a distance from the operator.

In some cases, the external distributing system will be the determining factor for the location of the switchboard, in order to have short internal conductors and reduce internal losses; where the current flow is large, the matter of conductor lengths becomes a very important factor.

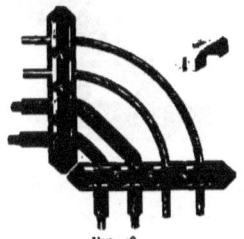
FIG. 38

There is another marked general tendency in switchboard construction, that is, to make the whole structure absolutely fire-proof; there is no reason why a central station should contain any combustible material other than the necessary waste and oil.

CONSTRUCTION DETAILS

There is at the command of the electrical engineer to-day fireproof material for every form of switchboard and insulation, necessary in central station construction. Asbestos affords a very good fire-proof covering for the conductors, and prevents them from conducting fire to different parts of the building; low-tension conductors may be bare, and supported by porcelain insulators (see Fig. 38) or bus bars supported by marble, as shown in Figs. 39 and 40.

Fig. 39

Every advance advocated by insurance inspectors in this direction is mutually important for the central station manager to preserve this class of property from destruction by fire, besides being an economical investment in the way of obtaining lower insurance rates on this class of risks.

Do not place the switchboard under steam pipes, or have an exposed window open on the back of the board. Do not allow automatic sprinklers to be placed over the board, for if they should act, the current leakage through the wet surfaces would cause a far worse hazard than could normally exist with a properly constructed switchboard.

The switchboard should be supported away from the wall, in order to have the back connections accessible; the underwriters' rules require such placement in several districts.

Fig. 40

The distance between the wall and the back of the board has not been sufficient in the larger boards, as there should be three and one-half feet in the clear in order that the attendant can properly work behind the board, and not make false connections with his tools.

FRAMING

Wooden frames, known as the skeleton form of construction,  have been used for the supporting of switchboard instruments. This step was taken to reduce the amount of combustible material used, and to separate the switchboard from the paneling or woodwork of the station. Fig. 41 shows method of joining and making skeleton switchboards.

FIG. 41

Oak or ash is the material generally used. The horizontal slats are placed at such distances apart that the apparatus can be readily secured to them. This form of construction was very much in vogue some years ago, but, on account of the fire hazard, it was abandoned where switchboard construction was seriously considered.

FIG. 42

The next step was to substitute slate for the slats, and still use the wooden vertical supports, with the instruments and devices mounted on the slate.

Later, I-beams, channel bars or "L" iron were substituted for the wood to support the marble or slate; in this way the space occupied by the supports for the board and bus bars was much less, and the clearances behind the board were more favorable to make good connections.

FIG. 43

Figs. 42 and 43 show methods of securing marble or slate to the verticals.

Standard steel sections being used, the switchboard can be readily erected; but the precaution to  have them insulated from the building structure should always be

FIG. 44

FRAMING AND CONNECTIONS

borne in mind where iron framework is used. This can be readily done by supplying foot-plates of marble, as shown in Fig. 44, and having the guys or expansion bolts, which stay it from the wall, insulated by a coupling, as shown in Fig. 45.

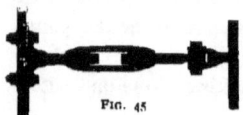

Fig. 45

In railway, high potential, and three-wire systems with grounded neutral, it is very essential to have the iron framework carefully insulated from the building structure, in order to bring up the ground resistance, as well as to prevent any jumping of current or running discharges behind the board to ground; also the hazard of injury to attendants working behind a switchboard which is insulated from the ground is greatly reduced.

CONNECTIONS

Methods of making connections between conductors behind a switchboard are of various kinds, depending on the form and purpose of the conductor. These various methods are connected together and shown in Plate 46; some of them are standard, and some have inherent weaknesses which have led to their abandonment.

The first shown is the familiar wrapped splice, which is used with bare conductors, where they are both served with copper wire and soldered together.

The second is the sleeve, where a thin brass tube is slipped over both ends to be connected and soldered.

The third form is a clamp connection, where both conductors are parallel and clamped together in one connector.

The fourth connection is where the wire is turned under the head of the bolt and screwed down.

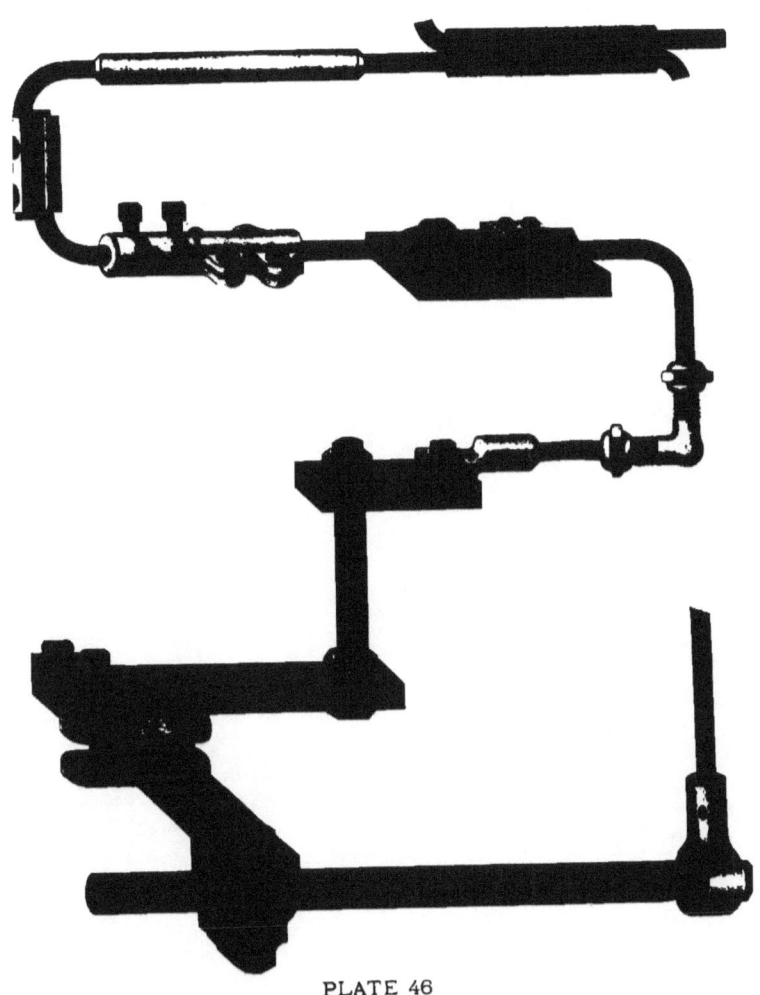

PLATE 46

CONNECTIONS

The clamp connection shown in the fifth is the "V" type, where two pieces with "V" recesses clamp the wire.

The sixth is an obsolete connection, where a sleeve is split and provided with a taper thread on the outside, over which screws a taper nut, and compresses the sleeve over the copper rod. The weakness in this connection consists of the fact that when the conductor heats, it expands and stretches the clamping nut so as to loosen the connection. All connections which have in their inception the surrounding of a solid conductor with a sleeve which is not strong enough to resist stretching under expansion, will eventually give trouble by heating.

The seventh shows the regular lug connection, and the eighth a connection often provided for in the terminals of back-connected switches; if the stud screws into the terminal and the nut locks these together, the arrangement is satisfactory; but if the stud passes through the switch with a nut only on top, and the bearing on the bottom a small shoulder, this method of connecting will eventually give trouble.

The ninth form shows where a threaded stud is secured to a bus bar by means of two nuts, which form a good connection. Angles are usually formed in bus bars by means of bolts; steel or iron bolts should always be used for this purpose.

The tenth form is the method used when a bus bar connects to a bus rod.

The eleventh connection is when the end of this rod is cut to a taper of twenty degrees, and a female taper lug is bolted down on it. Where these surfaces are ground together, it makes a very good form.

The twelfth shows the German method of securing a cable to a

CONNECTION EFFICIENCIES

lug, by forcing taper screws into the stranding, and in this way expanding the cable and securing contact.

There are a great number of other methods used, of connections for special purposes, but not of general application to switchboard connections.

We have in the case of a contact an effective negative coefficient for temperature; we may explain this in this way:

As the contacts expand, they tend to present more surface of contact, and are under contact at a higher pressure than when at normal temperatures. The effect of the increased efficiency of a joint at elevated temperatures is very clearly shown when the parts in contact are held by a steel bolt. Either brass or copper expands faster than the iron bolt, and under these conditions you can enormously increase the pressure on the contact and decrease the losses at this joint. The conductivity of the iron bolt is not of as much importance as this increased pressure by unequal expansion of the different parts of the conductor.

To take the volts drop on a connection, and multiply it by the ampères passing, will give the watts at that particular current density; but where it is a matter of contact surfaces, the drop will not follow as quickly as a current rises. To assume that this is proportional leads us into very grave errors.

In specifying any system of conductors for switchboard and operating devices, the losses should be expressed at full load, in volts drop.

BUS CONDUCTORS

There has been an erroneous idea that by laminating and allowing large radiating surfaces, conductors can in this way be kept cool, and consequently the losses reduced; some have advocated forcing the density up to as high as three thousand ampères per square inch for copper. The supposed gain is keeping the temperature down so that the resistance will not increase, due to the temperature coefficient for that conductor; but there are constant losses in energy, which, if saved by using better proportioned conductors, would pay a handsome interest on the investment for the additional copper. An example will illustrate this fallacy more forcibly:

Suppose we had six thousand ampères to carry one thousand hours in one year 40 feet. With bus bars at a current density of eight hundred and fifty square mils per ampère, and using five bus bars in multiple, 2 x ½ inch x 40 feet. $R = .0000668$. The watts lost will be 2,404 per hour, or 2,404 kilowatt hours per year, which if produced at a cost of one cent per kilowatt hour, the loss will cost \$24.04 per year. This bus bar weighs 771 pounds, and will cost, erected, approximately \$308.00.

Take the same case as above, but using a density of three thousand ampères per square inch, or three hundred and thirty square mils per ampère, and we will increase the radiating surface by using a bus bar 2 x ⅛ inch x 40 feet, which will have a resistance $= .0001688$. The watts lost per hour will be six thousand, and the kilowatt hours per year, six thousand. The cost of production is one cent per kilowatt hour; this will make the lost cost \$60.00 per year. The weight of the bus bar is 155 pounds, and the cost to erect \$70.00; the difference in the losses is \$35.96, and the difference between the two investments is about \$238.00, or for the additional expenditure of \$238.00, which would be necessary in order to have the current density 850 square mils per ampère, this additional investment will earn fifteen per cent. by the economy in waste energy, effected by this additional expenditure in copper. Again, in the case cited, the laminated copper bus, one-eighth of an inch thick, has to dissipate .33 of a watt per square inch of surface, whereas the larger bus has to take care of only .2 of a watt per square inch of surface exposed. In this case the larger bus is working more advantageously regarding ultimate temperature obtained, and will increase resistance less, due to this rise in temperature, which will again be in favor of the larger bus bar.

This example is worked out for the reason that a great deal of engineering is done on what is known as the least first cost basis, regardless of what this extravagant economy costs in operation. Such cases are more clearly demonstrated by practical examples than by generalities.

The materials of electrical engineering, especially those used for conductors, are more often put in by faith than by test. The station manager who will have his boiler-plate tested, which will not represent more than one-fortieth of the capital invested in the plant, will neglect to have his copper tested, which will represent anywhere from thirty to sixty per cent. of the capital invested in the plant; yet poor conductivity in the copper distributing system may seriously affect the dividend which should accrue to the installation, and the current uselessly frittered away in heating the conductors.

It will be a wise plan, and should always be required, where there is any considerable investment of copper to be made, to have submitted by the manufacturer a sample piece of fixed dimensions, delivered and tested for its conductivity, and if it is to be used for overhead work, it should be also tested for tensile strength.

The conductivity of copper is seriously affected by the presence of other metals, even in very small quantities, especially arsenic and tin. The most important impurity which will appear if the copper is not properly handled in smelting is the sub-oxide of copper. The brittleness of electrolytic copper is generally due to the presence of copper hydride formed during deposition. The low conductivity of over-refined fused copper is due to the presence of carbide of copper, which is formed in the presence of carbon as soon as the sub-oxide disappears. Copper impurities can only be detected first by the physical properties of the copper, and second by a chemical test.

The steel and iron manufacturers fill specifications requiring fixed physical properties in their products—so should the copper producer be required to fill both electrical and mechanical conditions, which are so important to the successful operation of a plant from a commercial standpoint.

Other conductors, such as iron, aluminium, etc., have been proposed, but in all cases the conductivity has been so low that the mass to carry any given current is from seven to thirteen times that for an equivalent copper bus bar, and this larger conductor requires much more space than can be afforded behind the switchboard; the insulating expense and the cost per unit of current carried is greater than with copper at the present market prices.

The dimensions of bus bars are generally selected by the current which they have to carry, and the connections which have to be made to them; two copper busses bolted together will carry about one hundred and eighty ampères per square inch of contact section, and the cross-section carries approximately twelve hundred ampères per square inch. Consequently the dimensions of the bus selected should be such that it will present proper area of contact for connections, without making them excessively long.

BUS CONDUCTORS

The current which is found in average practice, which can be carried economically by copper bus bars, is given in the table below; these current densities are covered by ordinary central station practices where the load factor is not greater than fifty per cent. The resistance per foot, the area in circular mils and square mils, and the weight per foot are given.

COPPER BAR DATA

SIZE	AMPÈRES	CIRCULAR MILS	SQUARE MILS	OHMS PER FOOT	WEIGHT PER FOOT
1 x ¼ in.	433	318310	250000	.0000336	.97
1¼ x ¼ "	530	397290	312000	.0000269	1.21
1½ x ¼ "	626	477465	375000	.0000223	1.45
1¾ x ¼ "	725	556400	437000	.0000192	1.70
1¼ x ⅜ "	676	596830	468750	.0000179	1.82
1½ x ⅜ "	798	716200	562500	.0000149	2.18
1¾ x ⅜ "	916	835600	656250	.0000128	2.54
2 x ⅜ "	1035	954930	750000	.0000112	2.92
2¼ x ⅜ "	1154	1074300	843750	.00000995	3.27
2½ x ½ "	1500	1591550	1250000	.00000672	4.86
2½ x ⅝ "	1715	1989440	1562500	.00000537	6.07
2 x ½ "	1222	1273240	1000000	.00000840	3.89
No. 0000 B. & S.	257	211600	......	.0000505	.64
½ in. Round	305	250000	......	.0000428	.76
⅝ " "	426	390625	......	.0000273	1.18
¾ " "	560	562500	......	.0000190	1.71
1 " "	861	1000000	......	.0000107	3.05

Bus bars are formed by having the copper billet drawn through graduated dies. If these dies are not properly graduated, or are dull, the surface of the copper becomes torn and reduces the usefulness for bus bar work, as a good connection cannot be made on an abraded surface. If the surface of the bus bar is warped it should be straightened, in order to make a good contact. Specifications for bus bars should be drawn so that they can be rejected for any of the above mechanical imperfections. Bus bars are drawn soft, medium or hard, as ordered. Medium is the bus bar usually ordered. Ordinary tools and ordinary tool speeds will work copper satisfactorily. If the tools tear instead of cutting the metal, use milk as a lubricant, and the cutting will be perfectly smooth.

COMPOSITIONS

In any mixture of other metals with copper, the resultant alloy will be reduced in conductivity below the mean of the different metals forming the mixture. There has been no law yet discovered between the conductivity of the different metals and the resultant alloy, from which the conductivity of the alloy can be predetermined. It will be seen in the table on following page that the conductivity of these alloys is more rapidly reduced when any quantities of tin are added to copper, than with an equal portion of zinc, except in the instance when tin is present in very small quantities, just sufficient to make the metal flow into a mould; this will give the highest conductivity of metal that casts readily, except in the case of one-half per cent. of silver and ninety-nine and one-half per cent. of copper, which gives a conductivity of eighty-nine per cent.

No old metals should be remelted where conductivity is required, as they have very low conductivities, and metal otherwise good may be burned in the pot and seriously affect the resultant com-

COMPOSITIONS

position. Brass is a mixture of two parts copper and one part zinc, and from the peculiar properties of this mixture, it is probable that there is a chemical combination at this ratio, and not a purely mechanical mixture, as with most alloys. So-called cast copper

COPPER PER CENT.	ZINC PER CENT.	TIN PER CENT.	CONDUCTIVITY COMPARED WITH COPPER AS 100 PER CENT. CONDUCTIVITY	VALUE AS A CONDUCTOR COMPARED WITH PURE COPPER AT 20 CTS. PER LB.
98.44	1.56	......	46.88	42.22 cents
94.49	5.51	......	33.32	59.20 "
88.89	11.11	......	25.50	78.40 "
86.67	13.33	......	30.90	64.60 "
82.54	17.50	......	29.20	68.40 "
75.00	25.00	......	22.08	81.40 "
73.30	36.70		22.27	89.80 "
67.74	32.26	......	25.40	78.60 "
00.00	100.00	.....	27.39	73.00 "
98.59		1.41	62.46	32.00 "
93.98	......	6.02	19.68	101.60 "
90.30		9.70	12.19	164.00 "
89.70	......	10.30	10.21	195.80 "
88.39	......	11.61	12.10	165.20 "
87.65	......	12.35	10.15	197.00 "
85.09	......	14.91	8.82	204.00 "
16.40	......	83.60	12.76	159.80 "
......	......	100.00	11.45	174.60 "

varies all the way from eighteen per cent. to eighty-nine per cent. conductivity of pure copper, and a number of methods are used, as well as mixtures, which will make the copper flow into a mould and conform to the pattern.

COMPOSITIONS

In switchboard construction the current-carrying appliances should be so designed that they can be worked directly out of pure copper stock; do not use cast metal where the current densities are high, and where the economy of material and labor is to be considered.

The bronzes are employed in special cases for their mechanical properties rather than their electrical properties, as their conductivities are low.

SWITCHBOARD MATERIAL

The actual surface on which the switchboard appliances are assembled should have the physical properties of possessing mechanical strength, insulating, and be fire-proof; besides, it is necessary that it be drilled readily and have a good surface on which to mount the different appliances, which are combined together to form the switchboard.

Wood possesses the mechanical strength and insulation when dry, but it is not fire-proof. There have been a number of attempts made to impregnate the wood with silicate of soda, or other fire-proof compounds, or even to paint it with fire-proof paint, in order to make it incombustible. Wood has also been covered with asbestos paper, in order to serve the same purpose, but none of these methods have proved satisfactory.

The natural stones offer the best material for switchboard surfaces, and, above all, slate is the most generally used. The insulating qualities of slate vary with different mines, and also with different parts of the same mine; but generally speaking, a uniform color, gray or brown slate, without any marked seams or veins, will be found to be a fair insulator. Slates which fracture readily along the veins, and in which the fractured surface shows a semi-metallic color, are always treacherous to use. Slate should have a bright, laminated fracture along the plane of its natural cleavage, and if it has a dull fracture, it is very liable to be too porous, and will absorb moisture, which will reduce its insulating value. As a rule, soapstone is too soft and fragile to be used for switchboard purposes, and does not hold enameling well. Slate surfaces are treated with enamel paint, and different imitations of marble and wood are in this way made, the enamel being burnt in and hardened and polished to a surface.

Among the marbles every variety can be found in regard to color and finish, but not many of them possess high insulating qualities. White Vermont is soft and fragile, as well as being too porous to exclude moisture, and a drop of oil will spoil a slab; for these reasons, it is very rarely used for switchboard purposes.

The granites are too hard to drill and holes will drift when hard spots are reached.

There are a number of Tennessee marbles, such as the gray and the champion pink, both of which have, as a rule, high insulating qualities and are easily worked; they also make very pretty combinations with dull coppered finished appliances.

Of the foreign stones, Italian marble and Mexican onyx make very good surfaces, and as a rule are good insulators. The former is probably more extensively used for switchboard construction than any other marble in America.

To drill marbles, use the slow-speeded twist drill; there will be no trouble in drilling marbles in this way, if the drill is not allowed to choke with the marble dust, because when this happens the drill gets hot and draws the temper. Where large holes are to be cut,

SWITCHBOARD MATERIAL

the quickest way is to use an iron pipe, the right size outside, with teeth cut on the lower edge; rotate this in the drill chuck and feed with emery; in this way the hole can be ground through very quickly. When there is trouble with the holes breaking out in the back, with large size drills, first drill an eighth of an inch centre hole and then drill with a large drill from both faces.

TILE CONSTRUCTION

Glazed tiles have been built up as a wall, forming a surface for the switchboard, the tiles being laid in cement, and the holes drilled through with hand drills, where required for securing apparatus; also cast-iron frames, in which are set porcelain blocks moulded with holes, so that apparatus can be thus secured and insulated.

SECURING AND MOUNTING SWITCHBOARDS

Natural stones, when secured to wooden structures, are very liable after a time to crack, by the warpage of their supports. In securing marble to iron structures, the only precaution necessary is to support the marble slab under the securing bolt, so that the marble is not sprung to conform to the surface of the iron work to which it is secured. Asbestos washers between the marble and iron, through which the securing bolts pass, form a very good bedding for marble slabs, and will support them permanently without danger of fracture. Natural stones can be ground and polished to within one thirty-second of an inch, and when so ordered, all slabs for one board can be cut from one block, so that all veining can be matched up, and the color will be uniform. It is necessary, in natural stones, to submit them to a test; the test usually made, where only low potentials are used, is by means of a magneto. There are two flat surfaces furnished with the leads to the magneto, and these are pressed on adjacent surfaces, or opposite surfaces of the marble or stone to be tested; if the magneto does not ring up with this connection, it is passed. All that this test indicates is that there is not a distinct metallic vein on the surface of the slate. It is the veins passing through the slate which form the conductors that render the slate useless as an insulator. The best way to test these veins is to use a sharp-pointed terminal, and stab the vein at two points close together; if, under this condition, you cannot ring the magneto, the veins are probably non-metallic. Where slate or any veined material is to be used for high potential, it should be submitted to the actual potential stress, and should require over fifty volts per mil potential before it breaks down; no marble should be exposed to potentials over six hundred volts, without the live terminals being separated from the marble itself, by micanite or other equivalent insulator, as marbles and slates leak excessively in moist weather, when they are acted on directly by high potentials. This leaking heats the marble, disintegrating it, and in some cases explodes it.

SWITCHBOARD APPLIANCES

OLTAGE measurements should be made with greater precision than any other electrical measurements in central station work, especially in low potential multiple arc distributing systems; a constant potential is necessary on the consumption circuits, in order to maintain an effective service and increase the life of incandescent lamps.

The conditions which surround measuring instruments in switchboard work must be taken into consideration in their selection. Those voltmeters having magnetic torque for their indication must be shielded from external magnetic influences when near iron structures or field magnets. This is generally taken care of in their design, either by shielding or working under such intense magnetic conditions that stray fields will not be of sufficient magnitude to introduce a commercial error.

With a varying voltage, it is essential that the instrument be dead beat in its action, and quickly respond to any impressed electro-motive force at its terminals.

The current flow through a voltmeter should be as small as possible, and the temperature coefficient of the resistance, in series with the moving system, should be low, and should give a negligible temperature error for the instrument to be of any real value.

An indicating instrument must be theoretically correct in principle, accurate throughout the whole range, simple in construction, and so proportioned to the work which it has to do that it will not be eternally getting out of order, and will stay in correct calibration.

To check up the accuracy of voltmeters, the temperature error is found by checking two voltmeters against each other when both of them are cold. Then place the voltmeter across the potential at which it operates for a number of hours, until its temperature has attained a maximum. Again check these two voltmeters, the cold against the hot voltmeter, and if the readings now differ from each other more than one per cent., the voltmeter should be rejected.

Another source of error in voltmeters is due to pivot friction, which is caused by the pivot wearing against the jewel bearing and abrading the surface, and increasing the friction between these two surfaces; this error should not be found in new instruments, as it is one which increases with the ageing of the instrument. In order to keep this friction as low as possible, the moving system should be light, and in order to test friction errors, the instrument should be moved quickly, in order to oscillate the index hand—note whether the hand always comes back and registers with the zero mark—current should be applied to the instrument, and different parts of the scale tested in the same way. The important parts of the scale where friction is found largest are near the standard readings, where it is continually used. The voltmeters should always be tested for the friction error in the same position in which they are used.

Means are generally provided on the switchboards for checking up the different voltmeters against each other, by means of plugs and a pressure switch. This should be done at stated intervals, in order to maintain the accuracy of these instruments, which are so important to the proper operation of the system.

SWITCHBOARD APPLIANCES

For alternating currents, the voltmeter should be calibrated at the same frequency as that on which it is used. In some types of voltmeters a change of frequency introduces a considerable error in alternating current instruments; especially where they are applied to circuits having a large power factor, they should be carefully checked for errors.

INSTRUMENT MOVEMENTS

The different principles employed in the movements of electrical instruments can, as a rule, be used for both volt and current measurements. The magnetizing force, in the case of the voltmeter, is supplied with a great number of turns of fine wire, and, in the case of the ammeter, is supplied with a few turns of coarse wire.

The oldest type of a measuring instrument was a permanent magnet pivoted over a conductor, which magnet was deflected by currents passing through the conductor, and these deflections were calibrated.

The movements of commercial instruments, as now used for switchboard work and commercial measurements, are only those which will be hereafter described.

The solenoid type, in which an iron core is sucked into convolutions, through which the current circulates, has a pointer attached to the core, and these deflections are calibrated. Fig. 47 shows an instrument employing the solenoid principle; in this case, the core is supported on a knife edge, counterbalanced by the pointer and an adjustable weight, and can be calibrated to give a fairly uniform scale for equal increments of current. This

FIG. 47

43

design is one which is used by the Brush Company on their arc circuits.

Fig. 48 shows another form of solenoid type; in this case the solenoid is circular and concentric to the curve core which swings around the axis of support, in response to current changes.

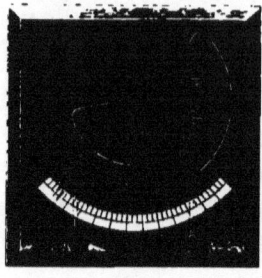
Fig. 48

By properly balancing this system by weights, a fairly uniform scale can be secured, or, for special purposes, the scale can be expanded at any point desired. This movement was used by the Edison Company.

These two movements are specially useful for current indications; they have an inherent error in measuring direct current, due to the iron core not responding readily to slight changes of current, which makes an increased reading lower than it should be and a decreased reading higher than it should be. This magnetic lag is called "hysteresis," and where there is any mass of iron employed in the moving system the accuracy is not sufficient for volt indications, except where alternating currents are measured, and hysteresis is, under this condition, not an appreciable error, but in this case, the precaution must be taken to laminate the core, so that local currents will not be induced in it.

Both of the systems above described are naturally heavy in their construction, so that the coefficient of friction of the instrument is large, and gives an error which is practically identical with hysteresis.

INSTRUMENT MOVEMENTS

Fig. 49 shows still another instrument of the solenoid type, which is used specially for alternating current work, being provided with a laminated core. This movement is used by the Westinghouse Company.

The perfecting of devices for the accurate commercial measurements of electrical currents and potentials is due to the untiring efforts of Mr. Edward Weston, who undertook, in 1880, to seriously develop for the electrical industries instruments by which the electrical quantities involved in electrical engineering could be accurately determined. After having experimented over the whole field of the possible forms of instruments, he determined on the type of

FIG. 49

instrument shown in Fig. 50, believing it to embody the best possible principles for the measuring systems of commercial instruments. This type is applicable to the measurement of direct currents; the

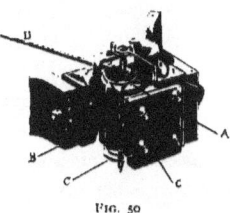

FIG. 50

measuring coil consists of a number of turns of fine wire, wound on a rectangular form, and this coil rotates in a concentric annular space, through which passes a permanent magnetic field. When current enters this coil, it tends to deflect it from a fixed position, which is reacted by differential springs; in this way deflections can be obtained, which are proportional to the current flowing.

"A" represents the internal iron core; "B" the moving coil; "C" the springs, and "D" the pointer.

INSTRUMENT MOVEMENTS

Where this instrument is used as a voltmeter, it has resistance in series with the moving coil. When used as an ammeter, it measures the difference of potential across a shunt, which difference varies with the current flowing. These instruments are made in a number of forms for switchboard work—the round dial type for isolated plants, illuminated dial type for central station work, and the edgewise type for heavy current work. All these are shielded by being closed in iron boxes, which form the case.

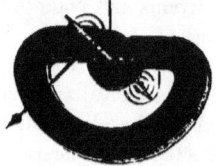

Fig. 51

Fig. 51 shows another instrument having the elements of a moving system rotating in a permanent magnetic field, this movement being restrained by springs which also carry the current into the moving systems. The instrument shown in this figure is known as the Kennelly Ammeter, manufactured by the Edison Manufacturing Company. This instrument consists of a flat permanent horseshoe magnet of semi-circular shape, with its poles brought out into semi-circles, and separated about one-sixteenth of an inch. In this narrow air-gap, and working in a strong magnetic field, is the disc armature "A." The windings are laid radially and symmetrically over the upper surface of the disc, and when the current passes through these radial lines, a magnetic pull is set up in the plane of the disc, which causes the armature to turn; the deflections are proportional to the current flowing through the system.

Fig. 52

Fig. 52 shows another method by which currents are measured by indirect means, due to the expansion of the conductor on being heated by the current passing through it and the expansion and contraction of this conductor being multiplied and actuating a pointer, which indications are calibrated. This figure shows several strands of wire, through which the current to be measured passes, one strand of which actuates the multiplying device, to which is attached the pointer. A disc is attached to the moving system, which enters a semi-circular air-tight box, and acts as an air dashpot, which reduces the oscillation of the meter and makes it approximately dead beat. This instrument is known as the Hoyt hot wire type.

The Cardew type of instrument has the conductor wound around the spindle of the indicating hand, and rotates this around as it expands or contracts; this movement is again calibrated. These instruments have special uses for shipboard work, as they are not affected in their indications by any external movement; twisted strips of different metals and spiral springs used torsionally have been used for the moving devices of measuring instruments.

FIG. 53

Fig. 53 shows what is known as the Wirt type, and consists of a magnetic field which is unsymmetrical to the moving iron system. The current is so disposed around the moving system that very little iron is used, and the hysteretic error, by careful treatment of the iron, is reduced to a negligible value. As the current flows through the loop surrounding the system, the iron tends to include a greater number of lines of force, and this moving effort is counteracted by weights or springs, so that the deflections are proportional to the current flowing. Where this system is used for a voltmeter, a

INSTRUMENT MOVEMENTS

number of turns wound on an elliptical spool, and the moving system placed concentrically to one of the curved ends, makes this form adaptable to volt indications. Hoffmann & Braun, Thomson-Houston and others use variations of the same principle for indicating movements.

ASTATIC VOLTMETERS

With high potentials there exists sufficient attraction between surfaces of dissimilar potentials so that a moving system can be operated by these surface charges.

FIG. 54

Fig. 54 shows a type, originally designed by Sir William Thomson, in which a moving vane enters between two charged surfaces; this system is pivoted and provided with adjustable weights, so that voltages can be read up to any potential which is only limited by the striking distance between the charged plates. This instrument requires no flow of current for the indications, and the scale is not proportional.

Other arrangements of the same principle have been made, one by Mr. Kennelly, where the system is suspended and consists of a horizontal vane, which carries vertical circular sectors of aluminum, which rotate in grooves formed by curved brass plates. Current is carried into a moving vane by a bi-filar suspension. When these adjacent surfaces are charged, they tend to rotate the vane proportional to the charging potentials.

FIG. 55

This last arrangement, on account of the fibre suspension, is hardly adaptable to switchboard construction. See Fig. 55 for details.

ASTATIC VOLTMETERS

The permanent magnet type of instruments is not suitable for the measurement of alternating currents. Mr. Weston devised the instrument shown in Fig. 56, which has, in common with his permanent magnet type of instrument, a rotating coil on which is wound the convolutions carrying the current to be measured, and the current is carried into the moving system by spiral springs. The magnetic field in which this system rotates is also formed by the current

Fig. 56

to be measured, and consists of two hollow helixes which enclose the moving system. There is no iron in this instrument, and it is not affected in its readings by changes in frequency. The oscillations of the meter can be damped by a brake which is depressed on pushing down the contact key. This instrument does not give a proportional scale, but one that is expanded in the middle.

RECORDING VOLTMETERS

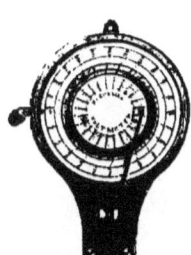

Fig. 57

In order that a permanent record may be kept of the variation of voltage on any system of potential, it is important that a visual continuous record be kept of the potential, and there are several instruments on the market for this purpose.

The one shown in Fig. 57 consists of a voltmeter movement of the solenoid type, in which the plunger is supported on springs, and has an index extension, on the end of which is a pen, by means of which a line of aniline ink can be traced on a dial.

RECORDING VOLTMETERS

This dial is moved around at a uniform rate by clockwork, and the variations of voltage are in this way registered permanently.

The Richard Brothers' indicator records in the same way as the above-described Bristol indicator, but the actuating mechanism consists of an iron vane attracted to the poles of an electro-magnet, and the pointer carrying the aniline ink is attached to this system.

VOLTMETER RELAYS

In order to control voltages where variable speeds are delivered to the dynamos, such as water-power or gas engines, it is necessary that an automatic method be used to regulate the generator thus driven, as compounding will not affect regulation for variable speeds.

A voltmeter relay is used for the purpose of actuating the rheostat which controls the field circuits. There are a number of methods by which this relay is operated, but generally by the solenoid and plunger principle, the solenoid being wound with fine wire, which is placed across the potential to be controlled. The plunger is restrained by a calibrated spring, and makes connection with one contact when the voltage is high, and with another contact when the voltage is low. A relay can be constructed so as to regulate within one-half of a volt.

The circuit thus controlled by the plunger may operate through the means of a motor or electro-magnet; the arm of the rheostat, cutting in or out resistance in the field circuit, maintains, in this way, a constant potential at the brushes of the generator. Such relays are also used to light signal lamps on the switchboard, to call the attention of the attendant to the fluctuation of voltage from the fixed standard.

VOLTMETER RELAYS

The central station in Paris, France, has its potential output almost entirely operated by automatic devices, which are controlled by automatic relays.

COMPARATIVE PRESSURE INDICATORS

In feeder distribution where the network of mains are supplied by feeders whose terminal pressure should be the same throughout the system, the pressure wires are brought back from the feeder ends and can be directly placed across the voltmeter and the potential measured, or the difference of potential may be compared against the selected standard feeder, by means of a comparative pressure indicator. The construction of this indicator is one having a permanent magnet needle pivoted, and is actuated by one solenoid across the standard feeder and one solenoid across the feeder pressure, to which it is connected; the effect on the magnetic system is differential, and the potential read on the instrument is the difference of potential that exists at the terminal of the measured feeder from that of the potential at the terminal of the standard feeder.

INDICATING WATTMETERS

In order to determine the power on any electrical circuit, it is necessary to multiply the current flowing by the volts pressure. An instrument which records this product is shown in Fig. 58. The main current flows around the moving system, and the potential on this circuit flows through the moving system, and the inter-reaction of these two forces is calibrated, according to their mutual forces, directly into a scale of watts which is the power element of the circuit being measured.

FIG. 58

INDICATING WATTMETERS

This instrument has special uses in the low potential systems, where the load curve in ampères does not indicate the power output of the station; for as the current consumption rises, the external distribution losses increase, which is compensated for by raising the bus pressure, so that an indicating wattmeter is the only indication of output which, when plotted on a curve, is comparable to the station's economical performance.

INTEGRATING WATTMETERS

In any variable load delivery, to multiply the current by the volts for a limited number of observations during the day, in order to determine the watt output, never leads to reliable results being obtained, unless, of course, the plant happens to be operating under constant load. For these measurements instruments of the integrating type have been devised, which operate by the combined effect of the current flow and potential difference; their rotations vary with the product of these two forces.

Their construction is, in all cases, quite similar to an ordinary shunt motor, the conditions of field and armature being reversed. The field is in series with the load; the armature, which is of high resistance and generally of Siemens' construction, with a commutator of a few segments, is placed "in shunt across the line," the shunt being taken off beyond the field coils. An outside resistance is placed in this shunt circuit, to reduce the current in the armature to the necessary small quantity, and to prevent any appreciable waste across the line. Also, a few shunt turns in some types act accumulatively with the series turns, to compensate for friction of armature rotation.

INTEGRATING WATTMETERS

Motor meters, as a class, rotate far more rapidly than is allowable in practice. It has been necessary, therefore, to introduce a "drag" or resistance to rotation, to slow the meter to a reasonable speed. This has been done in several ways, and perhaps the most common method is to attach a number of air fans to the shaft. These are quite largely used, and with fair success.

The resistance of an air fan to rotation is approximately proportional to the square of the speed; therefore, this device is only fitted for combination with such meters as have a torque increasing with the square of the current. However, since the torque of such meters does not quite reach the square, the retarding effect increases rather too rapidly, and has a tendency, although not always pronounced, to cause the speed of the meter to fall off proportionately on high loads.

Another method of "drag," which has been used with some success, is the rotation of a small fan in a liquid—a method perhaps rather better than the previous one, since resistance to rotation falls below the square of the speed when the liquid itself begins to rotate. Much depends in this case upon the shape of the receptacle containing the fluid.

FIG. 59

A third method consists in rotating a small inefficient dynamo, generally a mere disc turning between permanent or electromagnets. This resistance is of course directly proportional to the speed, and therefore to the torque. Applying this "drag" friction solves the principal difficulty to contend with.

53

INTEGRATING WATTMETERS

Fig. 59 shows the type of Thomson integrating wattmeter, which is adapted for both alternating and direct watt indications, and made especially for switchboard work. The main current enters the meter through the "U" turn of copper, which produces the field in which the armature rotates. The potential across the armature varies with the voltage, and the damping effect is produced by a copper disc rotating in a permanent magnetic field. The summation of the revolutions are recorded on a dial.

Fig. 60 shows an integrating meter which is specially used to measure alternating current. This meter is essentially an induction meter. The entire current to be measured passes through the primary, and an alternating field of force is developed in the direction of the axis of that coil. At the same time an alternating current is induced into the secondary, and this induced current develops another field of force in the direction of the axis of the second coil, and therefore at an angle to the primary, which produces a resultant field, which is constantly shifting; this field acts on a disc, and the rotations are calibrated to correspond with the currents flowing, and the indications on the dial will be practically ampère hours. The retarding devices are air vanes.

FIG. 60

DYNAMO GALVANOMETERS

In the control of dynamos, so that they may be thrown into service together, a means must be provided so that it is known when the potential of the generator is the same as that of the system on which it is to be thrown.

In small switchboards, a voltmeter switch is provided so that it can be connected across the terminals of the dynamo, and its

DYNAMO GALVANOMETERS

potential adjusted until it is the same as the system to which it is to be connected. For this purpose, in the older central station practice, dynamo galvanometers were used which were connected across one pole of the dynamo switch, the other pole being closed; when the dynamo is at the same potential the galvanometer reads zero and the machine can be thrown in on the circuit.

Another zero method which is used for this purpose consists of a differential galvanometer, one winding of which is placed across the bus potential and the other across the open dynamo switch. When the potential on both windings is equal, the hand stands at zero, and the potential of the machine is right for throwing in.

The incipient objection to a zero instrument for this purpose is that it gives the same indications for no current as it does for the proper adjustment of the generator; this has led to mistakes being made by the switchboard attendants, which has resulted in these methods being abandoned.

Now, each generator is provided with a small direct reading voltmeter, which is calibrated in volts and is provided with a small pilot switch, where there are several potentials used, so that the voltmeter switch will indicate the proper positions in which to throw the main dynamo switches when the dynamo potential is equal to the bus potential.

DYNAMO REGULATORS

The regulation of dynamos requires a variation of the current strength flowing through the field circuits, in order that they may be thrown into service, and also that the loads may be properly divided between the generator when working in multiple arc, and the proper potential maintained on the consumption circuits. In

DYNAMO REGULATORS

shunt dynamos their output is varied by a resistance in series with the shunt field, which resistance can be varied to increase or decrease the current flowing through the field circuit. The mechanical device used for varying this field circuit primarily consists of a continuous resistance, from which taps are taken at various points and brought to contact blocks, over which the contact arm traverses, thus cutting out or in resistance in this circuit.

The ordinary form is shown in Fig. 61, which consists of a circle of segments; the current is introduced through the pivot of the lever, and thence to contact buttons, and through the resistance to the other terminal of the regulator. The fault with this device, as usually connected, is that if the lever contact leaves the contact buttons the field circuit is open; if this happens when the machine which is controlled is working in multiple with others, serious damage may result.

FIG. 61

Again, in large generators many thousand volts may be induced in the field coils on the rupture of the field circuit, tending to break down the field insulation. For this reason the permanent connection should be made in all such regulators from the point "A" to "B," so that the field circuit is always continuously through the regulator, even if the lever is detached.

In any case where a field circuit of a generator over 50 K. W. is broken, this should always be done slowly, and the arc drawn in order that a partial circuit is maintained through the arc, so as to prevent the discharge of the field from rising to abnormal potential. In general, the field switches are provided with field circuit contacts, and on withdrawing the switch it enters another set of contacts, which throws a non-inductive resistance across the field terminals, so that when the field circuit is broken the discharged potential is suppressed.

The Siemens-Halske Company use a carbon auxiliary contact device, which allows the field circuit to be gradually broken between carbon points automatically, and suppresses in this way a sudden surging of dangerous pressure from the field discharges.

There has been a tendency for small units especially, to make the dynamo regulator compact. This, in itself, is a very good feature; but with the same watts lost as the size decreases, the temperature which the regulator attains, increases and uses the wire nearer its fusing temperature, practically results in too many breakdowns.

A larger factor of safety should be made in dynamo regulators, and much more than has been allowed in the past. A limiting temperature of ninety degrees Centigrade for enamel or grid construction, and fifty degrees Centigrade for open spiral coil construction, should not be exceeded. This latter form of construction has been used largely in central station work, simply because the factor of safety has not been sufficient in other types of regulators.

The cost of a regulator does not exceed three per cent. of the cost of the generator it controls in units, of the size of generator used in central station work. To increase the factor of safety of the regulator strengthens the weakest part of the regulating system and greatly insures reliability of service.

In regard to the total resistance of a regulator, this should be such that the dynamos' potential can be reduced below the normal at no load, when separately excited at normal potential. This is an important point, for a number of manufacturers obtain their rheostat data from tests on generators which admit of their being

DYNAMO REGULATORS

self-excited, and with this connection it obviously takes much less resistance to control the generator than when separately excited. In units from one hundred kilowatts up, the field currents become of considerable magnitude, and the regulators to control them have

FIG. 62

to be of special construction in order to properly carry and control these currents. The contacts and the connections are larger, and generally switchboards employing units of this size necessitate compactness; consequently the dial form of regulator is not suitable.

Fig. 62 shows a regulator in which the contact clips are arranged in two parallel staggered steps with a sliding contact bridging them, which contact is moved by a pinion in the contact head meshing into a rack on the side of the regulator.

There is also another form, in which the steps are arranged around in two parallel quadrants, and so staggered that eighty changes of resistance can be effected by moving the regulator arm through forty degrees. Both of the above regulators are connected to their resistance in what is known as a loop connection, which, in effect, is the same as drawing a contact along two legs of a U-shaped resistance, and in both regulators the moving of the contact upward increases the potential or load on the machine.

FIG. 63

In the quadrant type of regulator, it is also necessary to open the field circuit, and to prevent this being opened accidentally when the machine is in operation, an interlocking

58

DYNAMO REGULATORS

device is used, which prevents the field switch being withdrawn or thrown in unless the regulating lever is at the lowest point (see Fig. 63).

AUTOMATIC FIELD REGULATORS

Where dynamos are driven at a variable speed, the shunt field has to be brought up for a drop in speed, and down for a rise in speed.

There are several automatic regulators which can assist regulation considerably, if the variations are not too sudden. They employ an automatic relay which is operated by a given change in voltage, which, in turn, closes a circuit through a solenoid or motor, which actuates the contact lever to throw in or take out resistance; in this way compensation is made for speed variations. The contact lever is usually provided with a dash-pot or other equivalent damping device to prevent hunting of the regulator and surging in the electro-motive force of the generator controlled. Compounding will not take care of speed variations in a dynamo.

PROTECTIVE DEVICES

PROTECTION from abnormal flows of current, in a circuit which has a limited current-carrying capacity, is necessary for the safety of the system and the structure which contains it.

In any electrical distribution, there is a fixed consumption circuit, and the conductors are proportioned to the current supplied, and will take care of all normal demands with a predetermined loss. However, as the conductor system depends upon insulation to keep the current in the conductors themselves, and to limit the current in quantity by the consumption devices in that circuit, if the insulation fails or the current demand becomes excessive, this circuit then should be automatically severed.

The proper method to be employed for the protection of such circuits depends on the potential of the circuit, the current supply methods, and the character of apparatus to be protected, and the different devices which are used for the purpose of protecting circuits each have their own special sphere of usefulness.

FUSES

Much has been written on this matter, and there is hardly another example in electrical engineering where so much has been done in the laboratory without comprehending the true practical function of the device and its inherent limitations.

FUSES

The rating of the fuse is the determination of six variables:—its specific resistance, its temperature coefficient, the cooling effect of the terminals, the conditions for dissipating heat by convection and radiation, the specific heat of the metal, and the latent heat required by the fuse metal for its volatilization; the two last have more direct bearing on the time element of the fuse than on its rating. In use, the aging effects on the fuse are found to be the oxidization due to the elevated temperature at which it is used, and also its molecular stability changing under the variations of temperature.

These variables have all to be considered in the rating of a fuse; and to make a fuse for a specific number of ampères, in order that it can be connected with any kind of terminals, be blown in any position and at any external temperature or condition of moisture, has obviously led to the present doubt as to the accuracy of the fuse. In fact, any fuse can be so arranged that by conditions external to the fuse itself, it can be made to carry continuously a current flow one hundred per cent. above its rating.

The usefulness of the fuse as an automatic device is only realized when adapted to that duty for which it was originally intended, namely, to protect the conductor system from injury. The fuse being made the weakest point in the circuit, it should have such reliability as to not allow such an excessive current to flow through the conductor, so as not to create a fire hazard or injure the insulation of the conductor. It would be much more practical to number fuses to correspond with the size of wire which they are designed to protect, and be of such rating that they will blow between those currents above which the wire is allowed normally to carry, and below that current which will injure the conductor or its insulation, and there is plenty of margin between these two current capacities to allow for all the variables which will alter the current-carrying capacity of the fuse.

To send out a fuse marked to blow at a given number of ampères, or even fuse wire sent on spools, rated in ampères has led to the belief that they could be accurately calibrated; but without fixing the conditions of the length and terminals and holder, they in practice do not give reliable service. It is a physical impossibility to be sure that results can be reached in practice between one hundred per cent. of each other; yet to protect the conductors from excessive heating, due to abnormal current flow, is within their legitimate use; but to suppose that by overloading a fuse ten or twenty per cent. that it will blow, is not within the original intention or the present possibilities of this device.

Determining the sphere of usefulness of the fuse is a simple question of cause and effect. On circuits of high or low potential, not carrying currents over thirty amperes in most instances, an effective service can be rendered by the fuse, and it will operate quickly enough to fulfil the ordinary demand of a circuit-rupturing device. At these capacities the resistance of the fuse can be considerable without undue loss, as the heating of the fuse increases as the square of the current flowing multiplied by the resistance of the fuse; also the thermal conduction of the heat and the areas of radiation are low, and the mass of metal to be volatilized small; so fuses up to these capacities can be given a fairly reliable rating, and are in some measure independent of the variables, which assume large proportions in fuses at higher current-carrying capacities.

On low potential systems fuses are used extensively for the reason that they can be of such capacity that they will not operate unless there is an extraordinary demand on the system, for the current in these systems rises to an abnormal amount when the conductors are either grounded or crossed; but their use is being abandoned in the large network of underground mains from central lighting stations and copper strips substituted. It is considered better practice to-day to maintain current when a short-circuit occurs on these mains until it is burned out, rather than hazard the continuity of the service due to fuses blowing.

Fuses should not be considered as a reliable protective device, as it is a very important fact that, when the time arises for their action, it should be prompt; for it is evident that, by continuing the condition of short-circuit, excessive strains may be brought on the current-generating machinery and engine, as this strain has to be maintained long enough to raise the temperature of the fuse to the melting point and then supply sufficient energy to volatilize the fuse before the circuit is ruptured. Modern practice indicates that more than the external circuit itself has to be considered, as the reaction on the generating system, which is supplying the service, becomes a very important factor. In compound generators, where the effect of sudden rise in current demand on the generators is accumulative, it makes it very necessary that the current be severed quickly, as soon as it has reached a fixed value.

In modern central station practice the promptness of action of an automatic circuit-rupturing device becomes its most important function, where protection is to be afforded to the generating apparatus. The distributing system, however, can be subjected to a much greater variation in current flow, and stand the strain for a longer period; so that in this class of protection the time element of circuit-breaking is not so important. The fuse generally consists of a metal having a low fusing point, the combination of tin, lead and bismuth being those usually employed in the alloy. These are soldered to hard end terminals, usually of copper, the terminals having ears provided so that they may be clamped to the contact blocks. Fuses are also made for high potential service of a number of strands of fuse wire, each enclosed with a rubber tube. Also copper is used in the form of wire, but it has a very large time constant, and has to be maintained at considerable temperature if operated near its rated capacity. Copper fuses have also been used, wound in the form of a spiral, so that in rupturing the circuit the arc is drawn in a magnetic field, due to the convolutions of the fuse, and in this way extinguished. In order to reduce the radiating effect, fuses are enclosed in glass tubes, and in some cases are surrounded by a chemical which combines with the fuse metal itself on reaching a fixed temperature, and in this way reducing the resultant arc on the breaking of the circuit. In this same class are also fulminate fuses, which have, in close proximity to the fuse, a fulminate which explodes when the fuse reaches a certain temperature, and in this way ruptures the circuit.

AUTOMATIC CIRCUIT BREAKERS

A great many methods and devices have been employed for the purpose of mechanically rupturing a circuit by its own effect, after it has reached a predetermined point. The magnetic effect of the current itself is generally used to actuate an armature, which in turn releases the switching device. Expansion of conductors has also been used to disconnect the circuit, but the time constant of any heating effect, caused by the current itself, is far too slow to be effective, as is shown in the case of fuses.

In order to accomplish the result desired in a circuit breaker, the time between which the abnormal current rises in the circuit to be controlled and the actual opening of the circuit should be as short as possible; in other words, the time of opening should become less and less in proportion as the strength of flow increases, for the reason that the circuit which it controls may be short-circuited, and this current can quickly assume large proportions. It is important that any circuit-rupturing device should act as quickly as possible after the current has commenced to rise above the set value, so that the rupturing device can break the circuit before the current flow has attained such volume as would be destructive to any rupturing contact; and the quicker the circuit breaker acts, the greater the advantage to both the generating apparatus and the distributing system.

The actual breaking of the circuit should be performed positively and with certainty; yet to sever a high potential would necessitate drawing an arc which would seriously interfere with the proper action of the contact surfaces. Provision must be made so that this arc will occur where it cannot damage the current-carrying parts of the breaking mechanism.

THE CUTTER CIRCUIT BREAKER

This circuit breaker is the successful issue of many years' extensive research into the problem of automatically severing of electric circuits, and the perfected device as we have it to-day is the outcome of tests made under practical conditions, rather than the product of laboratory experiment.

Plate "C" shows the construction of this circuit breaker in side view and in part section. The main current circulates around the solonoidal coil "B" and tends to draw into the solonoid the movable plunger "C." The initial position of this plunger in the solonoid is determined by the adjusting screw "M." When the current is sufficient to overcome the weight of the plunger it is drawn into the coil with constantly increasing velocity, due to intensified magnetic action, as the polar distance or air space is decreased. When nearing the upward limit of its travel, having acquired a high momentum, it impinges upon the trigger "N" through the medium of the push pin "E." The immediate result of this is the release of the switch arm by the displacement of the retaining catch "F." The upper projection "H" of the trigger "N" is thrust against the striker plate "K," thereby utilizing the energy of the current to start the movement of the switch arm. This movement is intensified and sustained beyond the point of final rupture between the switch contacts by the thrust of the spring "O," which is released from compression by the initial action of the trigger. Thus the contact arm is thrown away from the contact terminal, and the circuit is opened.

Plate "D" shows the disposition of the main contacts and of the auxiliary carbon contacts through which the current flows after the copper switch plates have been severed. Upon these carbon surfaces the current is finally ruptured. By this arrangement the metallic contacts are preserved from the deleterious effects of an arc, the circuit being finally ruptured between the auxiliary carbon

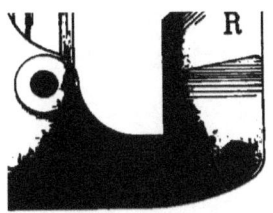

I-T-E
Circuit Breakers

CUTTER ELECTRICAL
AND MFG. CO.

1112 SANSOM ST.
PHILADELPHIA, U.S.A.

contacts. The efficiency of carbon for a final break is due to the fact that the vapor resultant upon the formation of an arc has a high resistance, and, owing to the refractory nature of this substance, but a relatively small volume is volatilized by the action of the arc, which, in addition, introduces into the partially severed circuit a counter electromotive force, approximating forty volts.

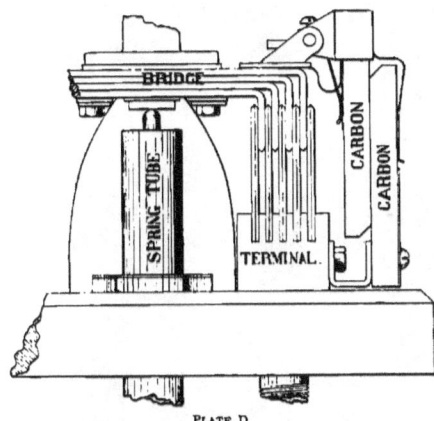
PLATE D

Among the important results attained by the development of the Cutter circuit breaker are:—First, the great reduction in the time elapsing between the occurrence of the excessive current and its final interruption. In the event of a fault or derangement in the distributing system the current tends to rise with extreme rapidity; the prompt interruption of the circuit, breaking the current before it has attained its full abnormal value, thus becomes a matter of the greatest importance. Some recent tests of these instruments show that the circuit breaker responded within five one-hundredths of a second after the occurrence of a short-circuit, and the current which would normally reach two hundred ampères was severed before eighty ampères flowed through the circuit breaker, showing that with normal inductance of gener-

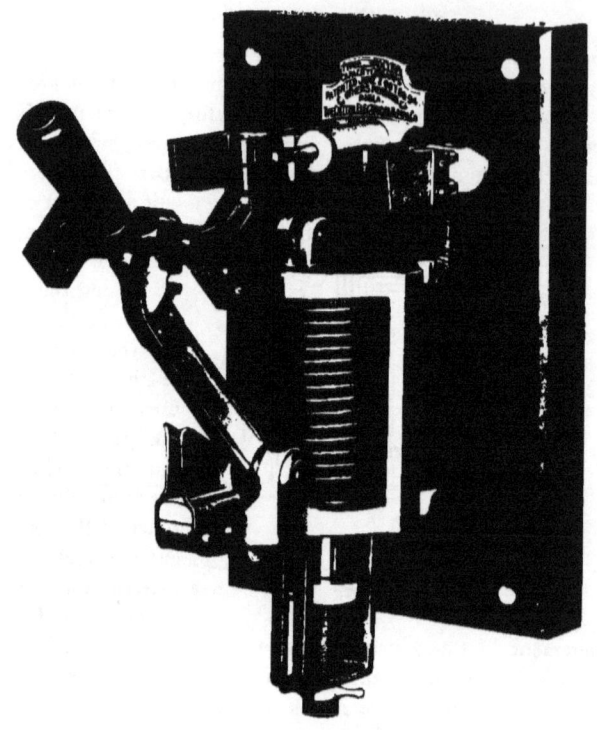

PLATE E
I T-E CIRCUIT BREAKER
STANDARD SWITCHBOARD TYPE.   SINGLE POLE.   5 TO 200 AMPÈRES

ating apparatus the circuit breaker acted more quickly than the generators could respond to the demand; consequently, abundant protection was afforded.

The responsiveness of the device upon the occurrence of a *gradual* overload is no less marked. Friction and the conformation of the magnetic circuit, if not properly taken care of in the design of a circuit breaker, will allow the current to creep beyond the point at which the device is adjusted to operate before the force of the solonoid is sufficient to actuate the plunger. In this instrument the plunger is free to move without appreciable friction, and, being worked far below the point of its magnetic saturation, it is extremely sensitive to current changes. This, in combination with the structural features alluded to, insures a positiveness of action which altogether precludes the possibility of "floating," as it has been termed. It may be hardly necessary to add that the circuit breaker is a very efficient protection against damage by lightning.

Plate "E" shows the standard switchboard type of the Cutter circuit breaker, of from five to two hundred ampères capacity, in an open position; Plate "F," a larger instrument of the same type, of from two hundred to twelve hundred and fifty ampères capacity, closed, while Plate "G" shows a still larger circuit breaker of standard switchboard type, single pole, of from fifteen hundred to three thousand ampères. All of these are intended for direct current only and are single pole, for use on circuits of six hundred volts or less, the long, clear break making them peculiarly adapted to the severe conditions incident upon street railway work. They are equally suited for lighting and power circuits, up to and including six hundred volts.

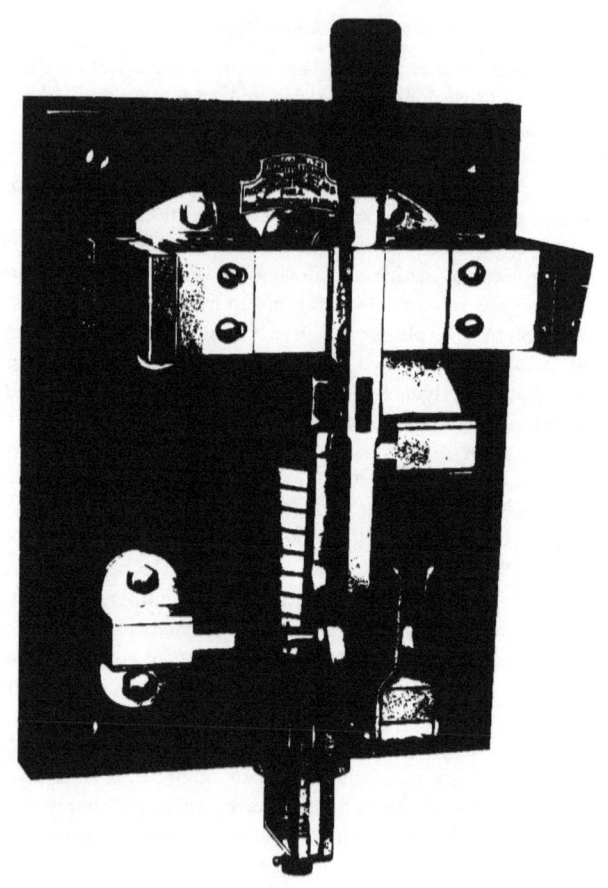

PLATE F
I-T-E CIRCUIT BREAKER
STANDARD SWITCHBOARD TYPE. SINGLE POLE. 200 TO 1,250 AMPÈRES

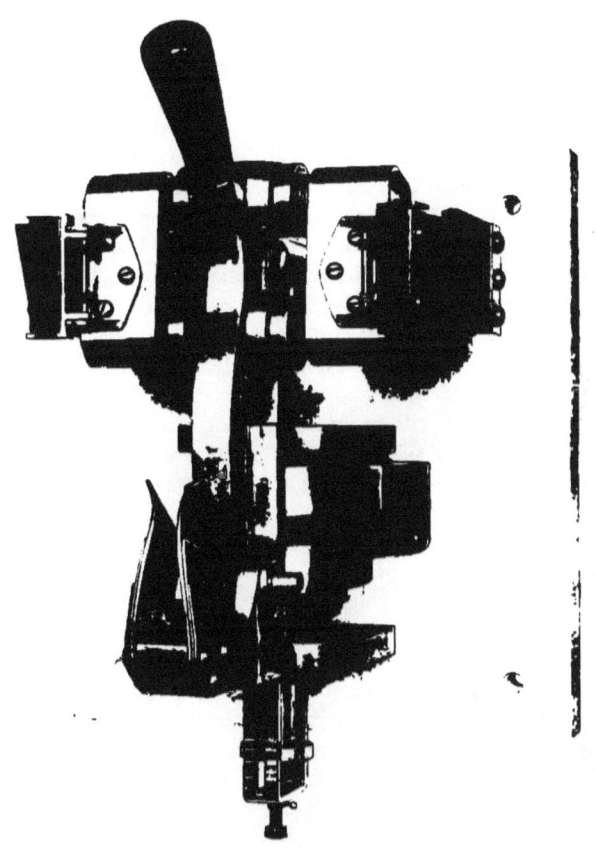

Plate G
I-T-E CIRCUIT BREAKER
Standard Switchboard Type. Single Pole. 1,500 to 3,000 Ampères

THE CUTTER CIRCUIT BREAKER

Plate "H" is a double pole circuit breaker, of a capacity of from thirty to two hundred ampères, for voltages of two hundred and fifty or less. As illustrated, it is intended for front connections, making it suitable for panel board work. For use on switchboards it would be used with back connections.

Plate "I" is a larger instrument of the same type, having a capacity of from two hundred to fifteen hundred ampères, double pole, for use upon circuits having a voltage of two hundred and fifty or less, equally adapted for lighting or power.

Plate "J" represents a circuit breaker of from two hundred to six hundred ampères capacity, double pole, double coil. Not only is this instrument designed to open both sides of the circuit, but, having two coils, it will be operated upon the occurrence of an overload upon either side as well as by an overload affecting both sides of the line, thereby insuring absolute protection of the circuit under any and all conditions.

Plate "K" represents a type of instrument especially designed to meet the severe condition of opening an alternating current circuit of two thousand volts or less. It is made in single pole only, and has a clean, wide double break of ten inches. In this instrument the coils of the smaller sizes are wound with covered magnet wire, and in the larger sizes the coils are made of open, bare rectangular copper. This type of instrument is made up to a capacity of two hundred ampères.

We have in the foregoing treated of the Cutter circuit breaker of an overload type only. With but a small increase in the size, and without in any way affecting the simplicity of the overload instrument, an underload function may be added. A principle analogous to that which insures certainty of action in the overload

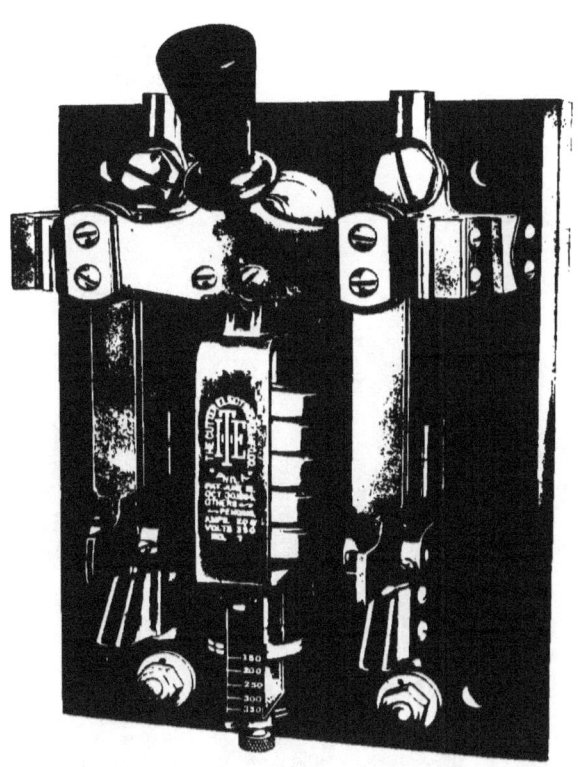

PLATE II
I-T-E CIRCUIT BREAKER
MIDGET SR. TYPE, DOUBLE POLE, 30 TO 200 AMPERES

PLATE I
I-T-E CIRCUIT BREAKER
STANDARD SWITCHBOARD TYPE. DOUBLE POLE. 250 TO 1,500 AMPÈRES

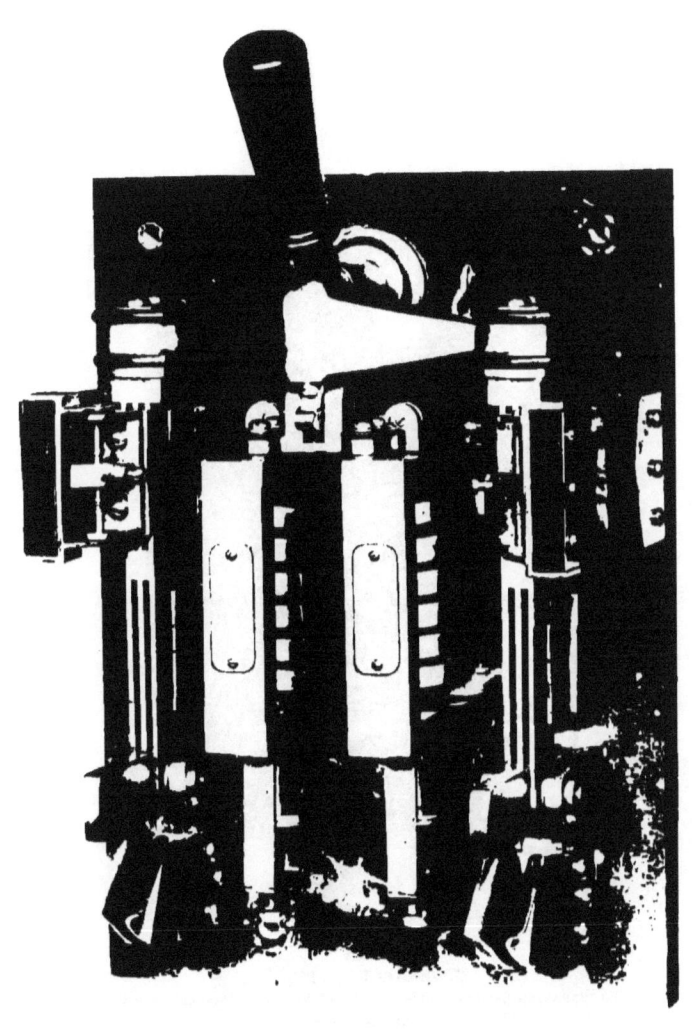

PLATE J
I·T·E CIRCUIT BREAKER
500 AMPÈRE. DOUBLE POLE. DOUBLE COIL

PLATE K
I-T-E CIRCUIT BREAKER
ALTERNATING CURRENT TYPE FOR HIGH VOLTAGE. SINGLE POLE ONLY
5 TO 200 AMPÈRES

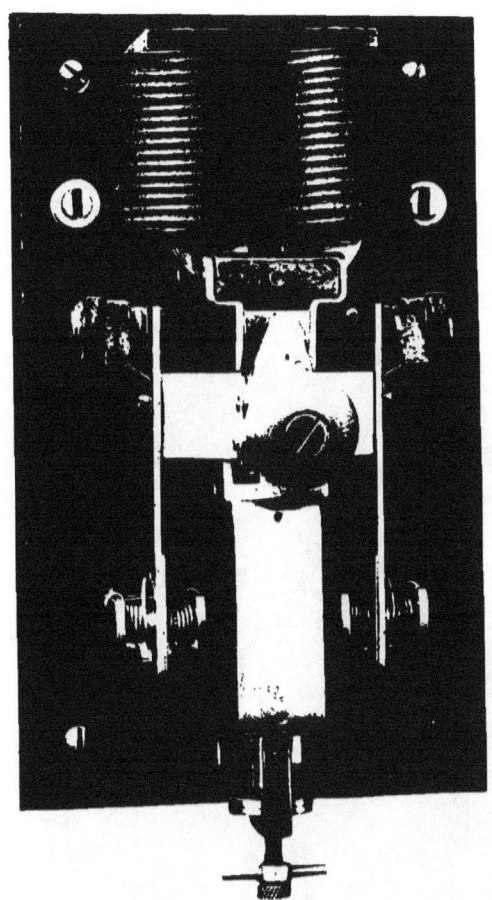

PLATE L.
I-T-E CIRCUIT BREAKER
DIRECT CURRENT, 110, 220 AND 500 VOLTS. 5 TO 25 AMPÈRES

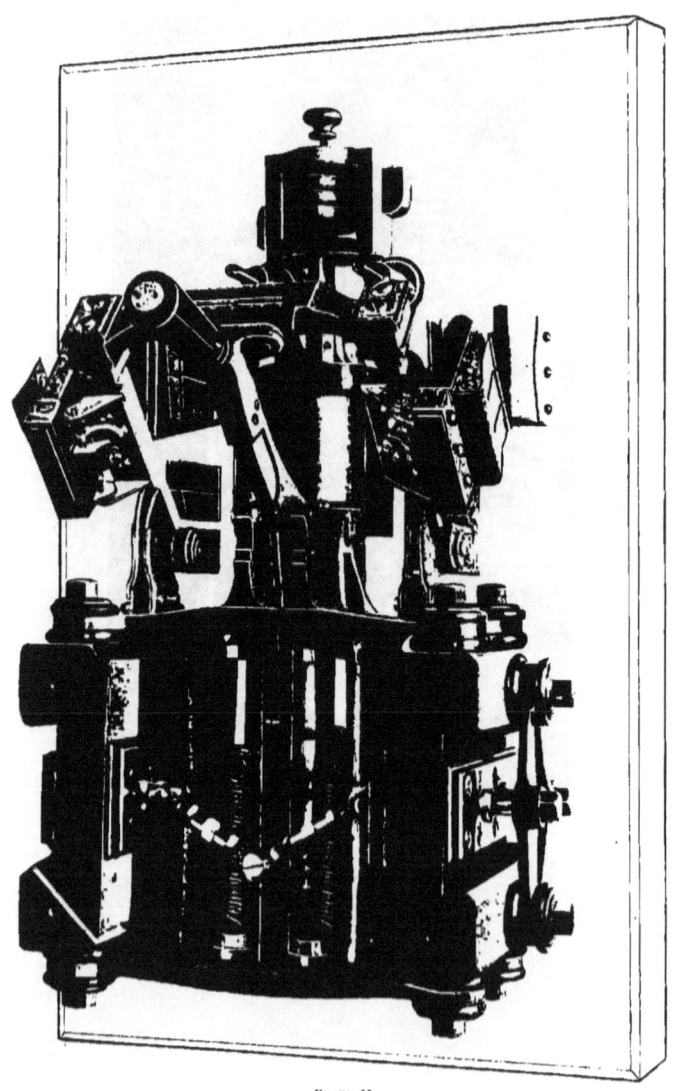

PLATE M
I-T-E CIRCUIT BREAKER
DIRECT CURRENT, EITHER SINGLE OR DOUBLE POLE. 4,000 TO 8,000 AMPÈRES

is made use of in the underload, the actuation of the trigger upon the occurrence of the underload being effected by the blow of an armature moving under spring pressure. It will be seen that whether the cause of operation be an underload, a "sneak current," or a heavy short-circuit, the trigger will be acted upon with a hammer blow, and never, in any case, without a free preliminary movement of the actuating body.

The underload device is made in two forms, one of which, operating only upon the interruption of the current supply, is especially suited for motor protection, while the second form, which operates upon the occurrence of a predetermined minimum flow, is peculiarly adapted for use in connection with storage batteries.

Plate "L" shows an overload circuit breaker having the underload function added. The type shown is intended for direct currents of from five to twenty-five ampères, single pole. This form is regarded as the best for storage battery protection.

Plate "M." This type of instrument is specially designed for circuits of very large capacity, either single or double pole. The construction is such that the current is carried through laminated contacts in series with the actuating coil of the instrument. The main break is between the laminated contacts, and is followed almost simultaneously by the auxiliary break, which is in shunt with the main contacts, and is protected by a final carbon break of ample capacity. All the features which have made the smaller types so successful are preserved in this device, while the arrangement of the laminated contacts and the general design are so accurately proportioned that the circuit breaker can be set with no greater effort than is required in closing a circuit breaker of fifty ampères.

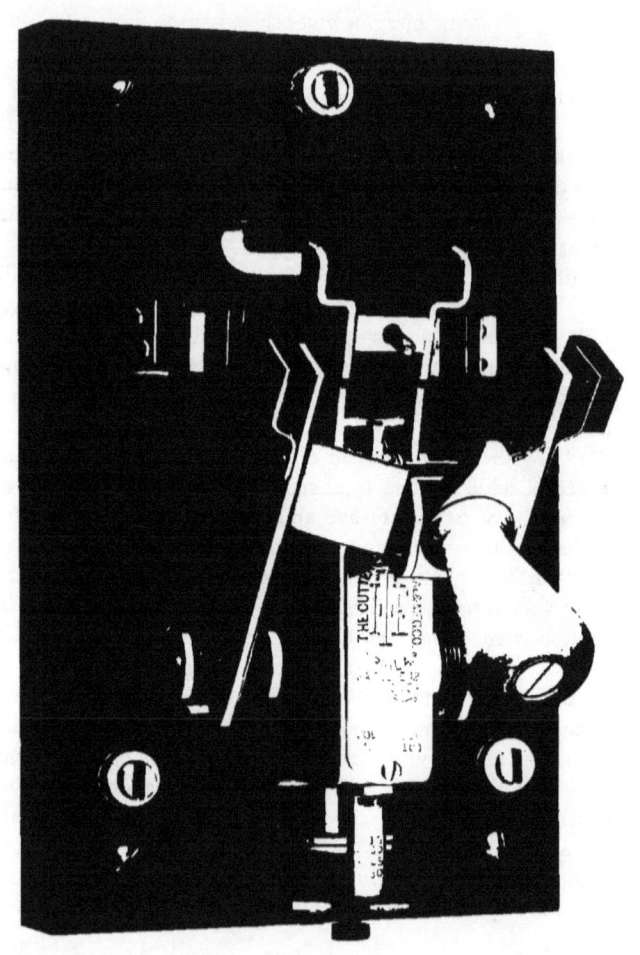

PLATE N
I-T-E CIRCUIT BREAKER
OVERLOAD AND "NO VOLTAGE." 5 TO 25 AMPERES

THE CUTTER CIRCUIT BREAKER

Plate "N" shows an overload circuit breaker having a "no voltage" function added. This type is intended for direct current of from five to twenty-five ampères, and is regarded as best for motor protection. They are made for one hundred and ten, two hundred and twenty, and also five hundred volts.

Plate "O." This instrument has been specially designed for use in connection with Edison three-wire currents, also for three-phase alternating current power circuits. This type of circuit breaker is made of a capacity from five to two hundred ampères, for use on circuits of two hundred and fifty volts or under.

I T-E CIRCUIT BREAKER
FOR USE ON CARS

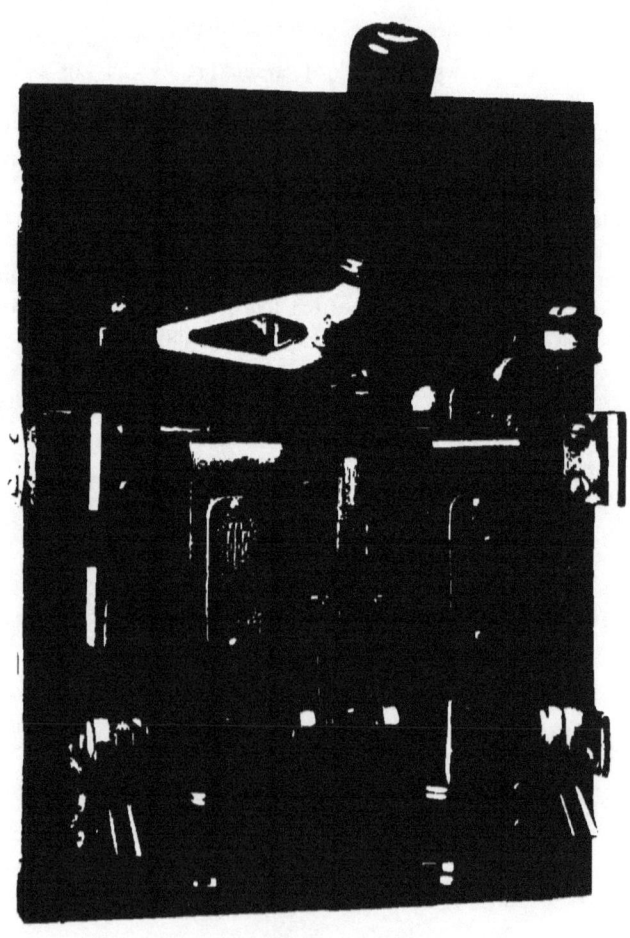

PLATE O
I-T-E CIRCUIT BREAKER
STANDARD SWITCHBOARD TYPE. TRIPLE POLE.

LIGHTNING ARRESTERS

IGHTNING arresters, in their function, bear the same relation to the potential stress of the station as the circuit breaker bears to the current demand, and their purpose is to offer to a high potential discharge a path to ground where it can pass off, rather than enter the station or the protected consumption devices, and break down their insulation in the effort of this discharge reaching the earth.

Lightning discharges have peculiar characteristics which make it comparatively easy to divert them. A discharge of lightning is supposed to be a rush of high potential which possesses an enormous frequency, and consequently inductive circuits arrest its flow; in this way discharges can be damped back from entering the station, and there will be an accumulation of potential adjacent to the inductive portion of this circuit. If an air gap be connected here, and one side connected to the charged line, and the other side of the gap connected to the ground, the discharge will jump this gap and pass to ground and be equalized, rather than force its way through the throttling inductive circuit. In order to accomplish this result, the conductor is wound into a number of convolutions, not over twenty, on a four-inch diameter mandrel, and these turns are interposed between the lightning arrester and the apparatus to be protected, preferably very near the lightning arrester itself. This protects the station from the discharge; but when the dis-

LIGHTNING ARRESTERS

charge passes the air gap, it so reduces the resistance of this gap, that when an active circuit is thus protected, the main current follows to ground and maintains an arc over the gap. For the purpose of again rupturing this circuit, several devices are used.

THE THOMSON-HOUSTON LIGHTNING ARRESTER

Fig. 78 shows type of magnetic lightning arrester. Under normal conditions the current from the generator passes through the electro-magnetic windings to the right hand wing of the arrester, and then to line, the left hand wing being connected to earth. The result of the current flowing through this electro-magnet produces

Fig. 78

a strong magnetic effect at the ends of the magnet, which is projected across the air gap; this forces the arc away along the curving wing edges, until it becomes too attenuated to be maintained by the machine potential. Practically this occurs instantly, and the arrester is then ready for another lightning stroke. Each arrester protects its own side of the line, and therefore two are required for each circuit. The illustration shows the form of arrester usually used on switchboards for central station protection.

WESTINGHOUSE TANK ARRESTER

The action of this arrester is to maintain an artificial ground of fairly high resistance on the lines to be protected, only when they are hazarded, and this type of leak arrester has been developed in order that there be no actual severance of the ground connection from the circuit to be protected; for in an arrester possessing an air gap, an abnormal potential must exist before the device operates. Here the systems to be protected are connected to a leak to ground during times of danger, and has electrically the effect of bringing the line to the level of the earth; for through this leak both line and earth are maintained at the same potential, and discharges pass off to ground through this leak. An inductive circuit is usually interposed between the apparatus to be protected and the external line and leak, in order to force this discharge to ground.

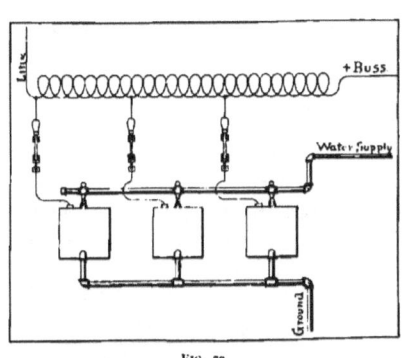

Fig. 79

A leak arrester allows the surging of induced potential in the line protected, due to the inductive effects of charged clouds over the line to follow the same potential values as the earth over which they are strung. The type of leak arrester is only suitable to protect circuits where one side of the circuit only is to be protected. Fig. 79 shows this type of arrester, and the tank leak is plugged when the system is threatened. There are generally three tanks employed, having between them inductive circuits, "A," "B," and "C," so that if the discharge passes one, it is restrained still further by the other two.

WESTINGHOUSE TANK ARRESTER

There are a great number of types of arresters using fuses with air gaps, the fuse being severed by the current following the discharge; but as discharges follow each other rapidly, continuous protection is desirable.

There are also condenser arresters in which the lightning discharge passes from one plate to the other; but these plates being in series, the initial potential of the circuit cannot maintain a discontinuous arc.

THE NON-ARCING LIGHTNING ARRESTER

It has been discovered that with certain metals, when they form the surfaces for a spark gap, an alternating current will not maintain an arc between these surfaces, due to the cooling effect of the terminals and the character of the vapor of the metal in the arc. Fig. 80 shows the type of the Westinghouse Non-Arcing Arrester, which is constructed embodying this principle, and which consists of a number of cylinders of non-arcing metal, forming several gaps between them, which are jumped by the high potential discharge; the alternating current does not follow and maintain an arc across these spaces or gaps, and this form of device is largely used in alternating current switchboards, where it is considered policy to place the arrester on the switchboard itself.

FIG. 80

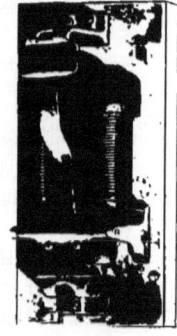

FIG. 81

THE NON-ARCING LIGHTNING ARRESTER

Yet another form of arrester is shown. In Fig. 81, in this case, the current, in its passage to the ground after following the discharge through the gaps, actuates an electro-magnet which pulls an arm, and lengthens the air gap until the arc is extinguished.

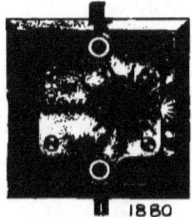

THE LOW TENSION SWITCHBOARD

IF the elements of design, common to both the isolated plant and the large low tension central station switchboard, both can be treated together. Taking the simpler forms first, and afterwards considering the special conditions necessary to be met by the switchboard, when used to control the more intricate methods of central station operation and distribution— methods recently introduced to give the proper potential supply over the expanding areas of the external network of conductors, and also for working of the storage battery in connection with the generators of the station.

The isolated plant usually consists of several generators and a number of feeders to centres of distribution. The actual assembly of the apparatus on the switchboard, and its relative position, is so largely determined by the local conditions to be met, and the convenience in handling, that to illustrate the different methods of assembly would not be of much value, but the method of connecting the different appliances is common to all the systems.

Diagram 1 gives the connections where a shunt dynamo is operated on a simple two-wire system. In this case the two terminals of the dynamo are brought to the switchboard through the dynamo leads to a double-pole switch.

The connections are usually made to the switch, so that the dynamo feeds through the fuses and then through the switch mechanism to the bus bar, so that if the switch is pulled, it can be fused when there is no current on the switch. Circuit breakers are usually inserted in the dynamo lead before it is tapped on the dynamo switch; the main current is here taken, one leg to the bus bar, and the other leg to the shunt dynamo ampèremeter "A," and then to the other bus bar. The dynamo regulator "R" is connected in series with the shunt field, the field wire being brought from the dynamo for this purpose; one end of the field being connected to one lead at the dynamo, and the other end connected to the other dynamo lead at the switchboard. The voltmeter "V" is connected across the bus bars, and the ground detector "G T L," in one-hundred-and-ten-volt system, consists of two lamps in series across the bus bar, a middle connection being taken between the two lamps to ground. Generally the ground connection is provided with a plug "P," so that the system is only grounded when the test is being made.

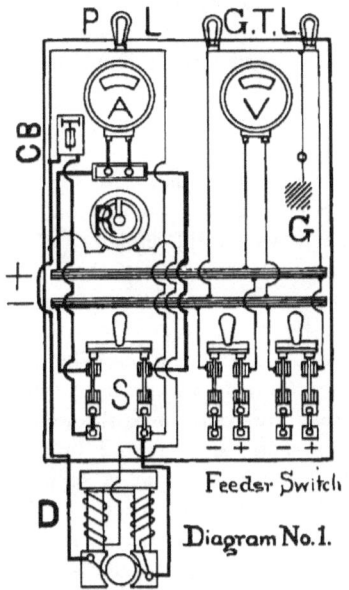

Feeder Switch
Diagram No. 1.

In connecting two or more dynamos that are to be worked in multiple, provision has to be made so that these dynamos can be thrown together without any fluctuation to the potential of the system.

Diagram 2, Fig. 1, shows a method where a dynamo galvanometer is used for this purpose, and is so connected that the dynamo feeds into the bus bar one side direct, and the other side through the dynamo galvanometer, which usually bridges the open dynamo switch, two single pole switches being used when this method is employed for throwing in dynamos.

It is evident that when no current flows, the potential on the dynamo and the bus bar must be the same, when the dynamo can be thrown on the system. Fig. 2 shows another zero method where the potential for the generator and the potential on the buses both act on the same magnetic system differentially; when the currents through them both are equal, the needle of the differential galvanometer "D G" will point to zero, and the dynamo is ready to be thrown in.

Both of these zero methods possess the inherent objection that they give the same indication when there is no current flowing through the instrument as when they are ready to be thrown in, and mistakes have arisen from this cause, and have led to voltmeters being used for this purpose. Each generator is supplied with a pair of contact buttons of a voltmeter switch "V S," which are connected direct to the two leads of the dynamo, and the voltmeter can be connected directly to any machine and its potential raised until it is the same as the bus, when it can be thrown in.

In compound generators, where more latitude can be allowed in the potentials of the generators being thrown in, pilot lamps "P L" may be sufficient to give the proper indication when they are con-

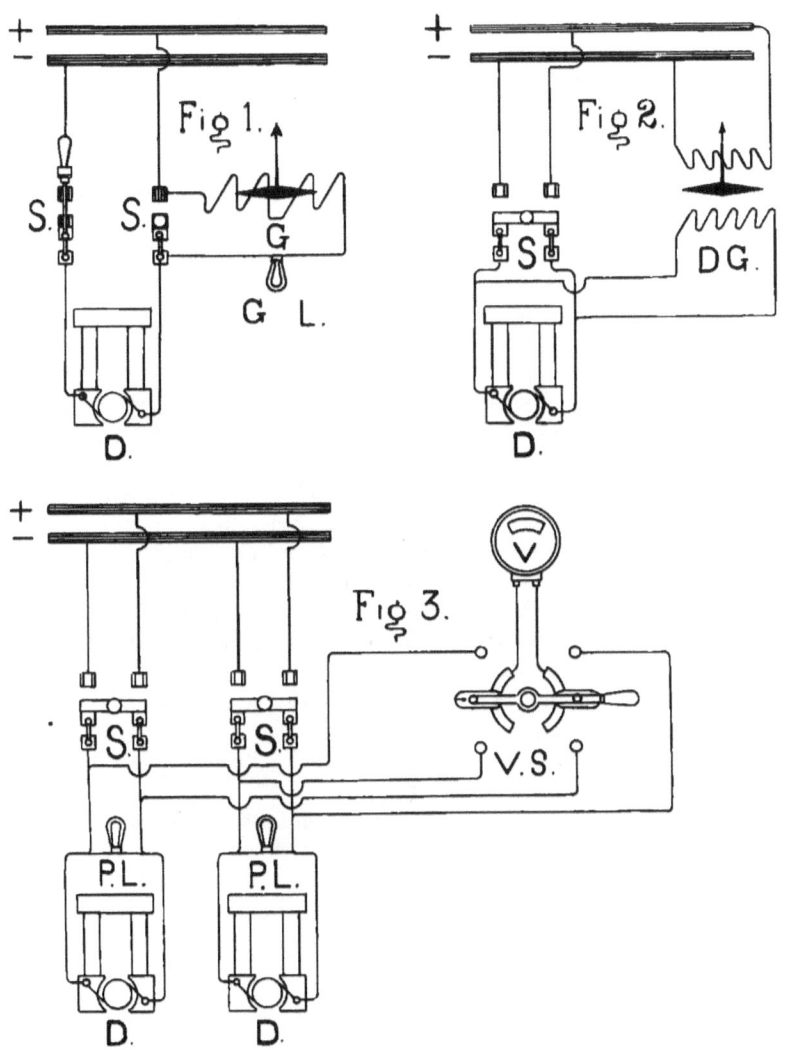

nected across the dynamo leads. These two last connections are shown in Diagram 2, Fig. 3.

The above connections show those used for shunt generators, where the rheostat is the only method of regulation; but isolated plants are usually provided with compound wound machines, and provision has to be made on the switchboard for equalizing connections, as it is the usual practice in isolated plants to have the equalizing bus on the switchboard.

The proper connection of the equalizer plays an important part in compound dynamo regulation, and the resistance of the leads from the dynamos to the switchboard, both equalizing and series leads, should be calculated as follows:

In all generators of the same size and having the same compounding characteristics, the resistance from the equalizing bus to the bus connected to the series side would be equal for all generators. Where these generators are different sizes, the drop of potential on the equalizer and series leads should be the same for all compound generators, when these generators are working together and carrying their full load. Where the dynamos are not of the same type nor the same compounding characteristics, the resistance of the equalizing circuit will have to be such that the drop from all generators will be the same when they take their maximum load together; but it is hardly possible to compound dynamos of very different characteristics together, so that they will pick up their proportional load equally among themselves, and only the best approximate condition can be obtained, when they will take their full load together.

If it is necessary to operate compound generators on two or more potentials, each potential must have its independent equalizing

THE LOW TENSION SWITCHBOARD

bus, so that the equalizing leads can be shifted to the equalizing bus to correspond with the potential bus, on which that dynamo is to be operated, in order that the proper regulation can be effected by the series field. Diagram 3 gives the connections for two compound generators connected in multiple. This diagram shows a three-pole switch used for this purpose. The third or middle clip should be made longer, so that the equalizing connection can be established before the generator is thrown in on the system. An equalizing bus is provided in this diagram, which is the usual practice, where there are a number of machines working together.

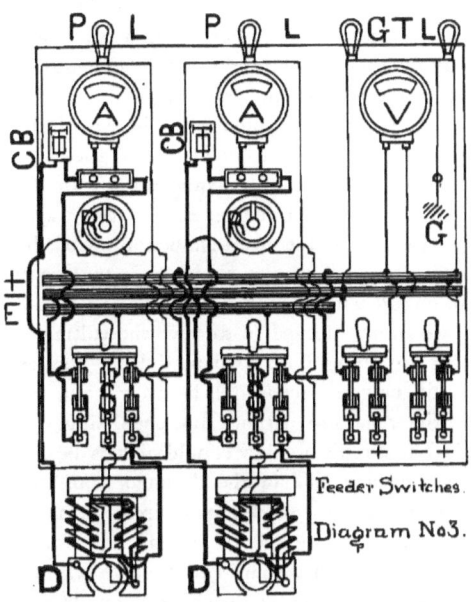

Diagram No.3.

Another method, which is sometimes used, is to connect the equalizer and also the dynamo lead from the equalizer side of the generator to a double pole switch. This can be thrown in and the dynamo brought up to potential, and a single pole switch can be closed on the open side of the generator to throw the machine in on the system. This method of connection has its advantage:—if there is much distribution drop, and if the generators are over-compounded to take care of it, and a number of machines work together in multiple to supply the full load when this maximum drop occurs, it is evident that when one machine is operated to supply the minimum load, the generator will normally give too high a potential, and hand regulation will be necessary in order to obtain the proper potential; whereas, if the series and compound connection were left in circuit of the idle machines, they would reduce the normal over-compounding of the active machine as the load on the plant decreased. In this way better regulation can be produced and the over-compounding of the machine is somewhat controlled. This also avoids ever putting in a compound machine in circuit without first equalizing.

Another method is to have the voltmeter by which the generators are brought up to potential, not connected across the machine until the equalizer switch is thrown, and in this way avoid putting in the generator without first equalizing.

The circuit breaker should always be connected in the lead on the opposite side of the dynamo from the series winding, for it would require a double pole circuit breaker to open the circuit on the compounding side of the machine. Diagram 4 shows the connection for a three-wire two-dynamo system, which, in effect, is two dynamos in series, the positive of one brush being connected to the negative of the next, which forms the neutral connection. Besides the apparatus required on the two-wire switchboard, the three-wire system requires two voltmeters, one on each side of the system, and the three-wire ground detector should have two lamps in series between the positive and negative bus to ground, and one lamp between neutral bus to ground; the ground detector should be provided with a double throw switch, so that the positive or nega-

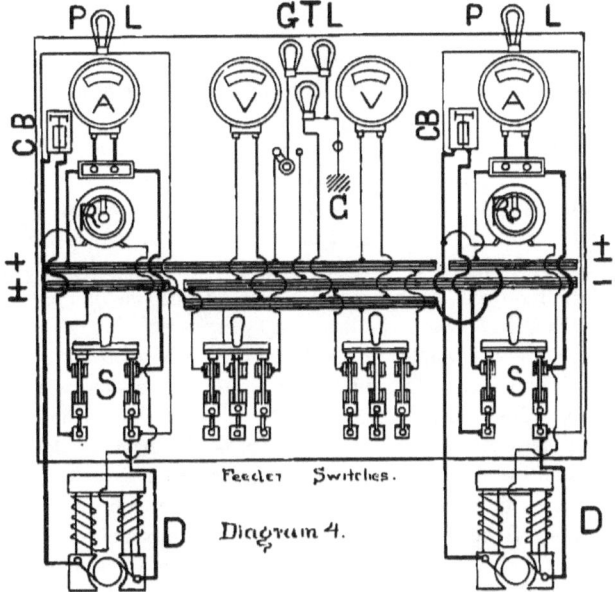

Diagram 4.

tive side of the system can be tested with the same detector, three-pole switches being used for all feeders.

Sometimes the neutral connections are made between the dynamos themselves, and the neutral bus is only used behind the feeder board and connected to the common dynamo neutral in the dynamo room. This saves nearly one-third of the copper in leads over that required to bring every terminal of the dynamo to the switchboard.

Where dynamos are to be used on either side of the system, they have to be provided with a double throw switch, so that they can be connected on either side, and the connections must be made so that the polarities will be right for either position of the switch.

Diagram 5 shows the connection for two sets of compound dynamos connected to a three-wire system. Here the only departure from Diagram 4 is that an equalizer has been added, one for each side of the system, and of course if these machines are to be used on either side of the system, a double throw three-pole switch would be required, so that the dynamos on one side of the system always equalize on the same bus.

There are a number of methods used where, from a single dynamo of two hundred and twenty volts, a three-wire system is operated. These systems require an auxiliary device which would equalize the loads between the two sides of the system. It is evident that if two dynamos were connected together in series across a 220-volt two-wire system, and the neutral taken from the common connection of these two dynamos, and if an unbalanced load were operated, the dynamo on the highly loaded side of the system would tend to operate as a dynamo and pump current into that side of the system, and balance the system external to the generator itself. The switchboard for this system, as far as the generators go, is simply a two-wire switchboard, but the equalizer or compensator is connected across from the positive to the negative, and the neutral taken from the common junction of these two to the feeder switches only.

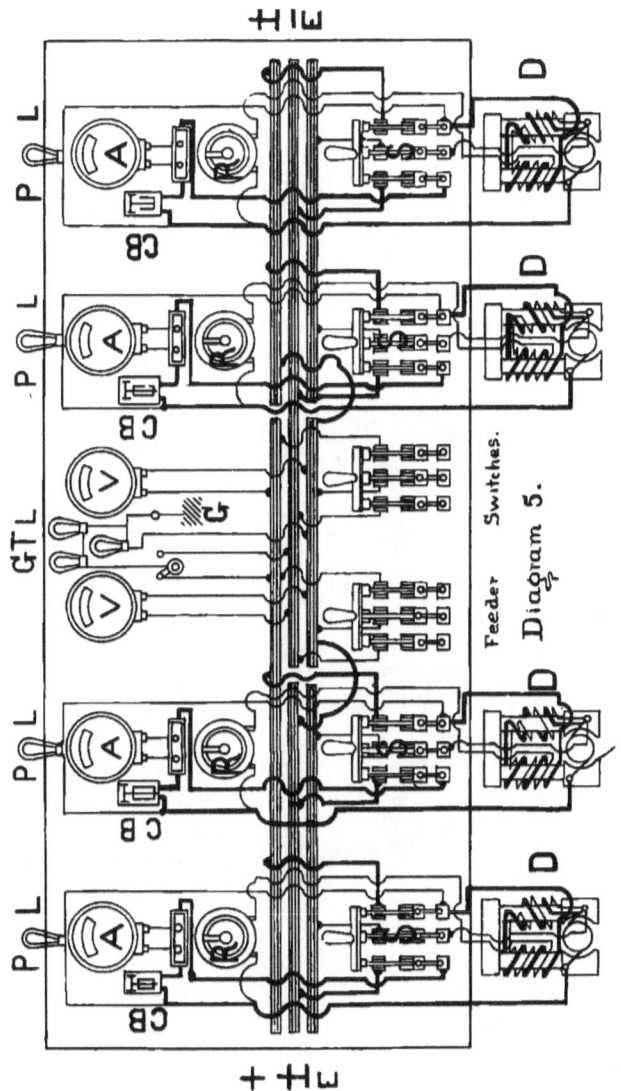

THE LOW TENSION SWITCHBOARD

A storage battery has also been proposed for affecting an equalization of the load between the two sides of the three-wire system, but the connections in this case are the same as an equalizer. A five-wire system is usually connected as a three-wire system, except in having four dynamos in series with connection between each generator, and the dynamos connected between the different buses at the back of the board. The connections for the above methods are shown in Diagram 6A.

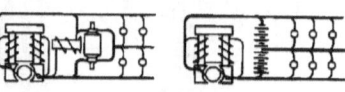

Diagram No 6 A.

It is often important, in isolated plant work, that the motors, especially if elevator motors be used during light loads, be operated on independent generators; these loads fluctuate violently and the generators cannot regulate quickly enough in order that the constant potential be kept on the lighting system.

The switchboard is separated up into two bus systems—lighting and power—and each generator is provided with a

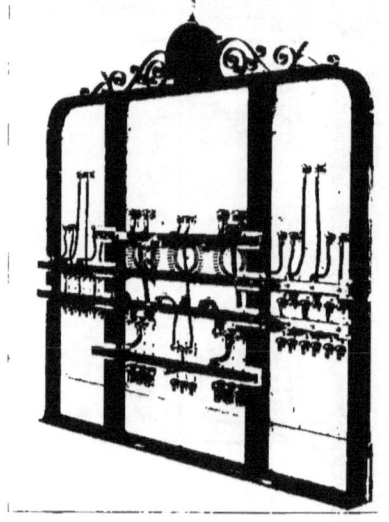

FIG. 82

double throw switch, operating on power in one position and lighting in another.

In large apartment houses, public buildings and asylums, it is also advisable to separate the public lighting from the private lighting, and supply it by the same or different generators. This is taken care of by providing the generators with double throw switches, and also supplying the feeders with double throw switches, if they are to be fed from different sources of supply.

Some methods of making back connections for isolated plant switchboards are shown in Fig. 82.

LOW TENSION CENTRAL STATION SWITCHBOARDS

EFINITE control of potential over a large network of low tension conductors has required a variety of methods in the handling of these feeders at the switchboard, in order to produce this variation of potential economically.

The first means resorted to in order to create this difference of potential on the feeder terminals was to insert an adjustable resistance to compensate for the unequal losses occurring in the feeders, so that the current would be delivered to the mains at a uniform potential throughout the system. This method, however, is uneconomical, both as regard the amount of energy consumed for the purpose of regulation and the valuable space occupied by this method.

Fig. B, of the introductory, shows this method with equalizers, as installed in the old Adams Street Station of the Chicago Edison Company. This method has now become practically obsolete; but as the low potential systems have been expanding over large areas to include a greater number of customers, it has again become necessary to deliver current at the station ends of these feeders at different potentials, to compensate for the unequal loading of the system. This is affected under some conditions by running the dynamos at different potentials, which supply independent busses, the feeders being so arranged that they can be thrown on any of these busses and supplied with current at the proper potential to make good the feeder losses and deliver their current to the mains at a uniform potential.

When the units in the station are small, they can be divided up on the different busses with comparatively high economy; but as the central station business has increased, and the plants have adopted large units, both on the score of economy for the first cost and operation and to increase the kilowatt output to the square foot of floor space, these practical conditions have greatly altered modern central station switchboard design. The condition of operating the units at an economical load must be maintained, while the feeder service requires several potentials to be delivered to it.

The booster system has been devised in order to change the potential of the current delivered by the generators "D D" in Diagram 6. In order to clearly understand the connections for the system, the following explanation is made regarding its action in the case of the three-potential system:

In this case, the medium potential bus is fed directly by the units operating the station, and the current is delivered from the medium bus to the high potential bus, through the armature of a dynamo whose field is separately excited, the current capacity of the armature being sufficient to carry the loads, at which it would be uneconomical for one of the units to operate. The current, in passing through this armature, has its initial potential raised, and the amount of this increased potential is controlled by a field regulator "R," in series with the dynamos' separately excited field; the currents which supply the bus of lower potential than the medium bus also flow through an armature, but in this case the initial potential of the current is reduced before it is delivered to the low potential bus. This dynamo is operating as a motor, and

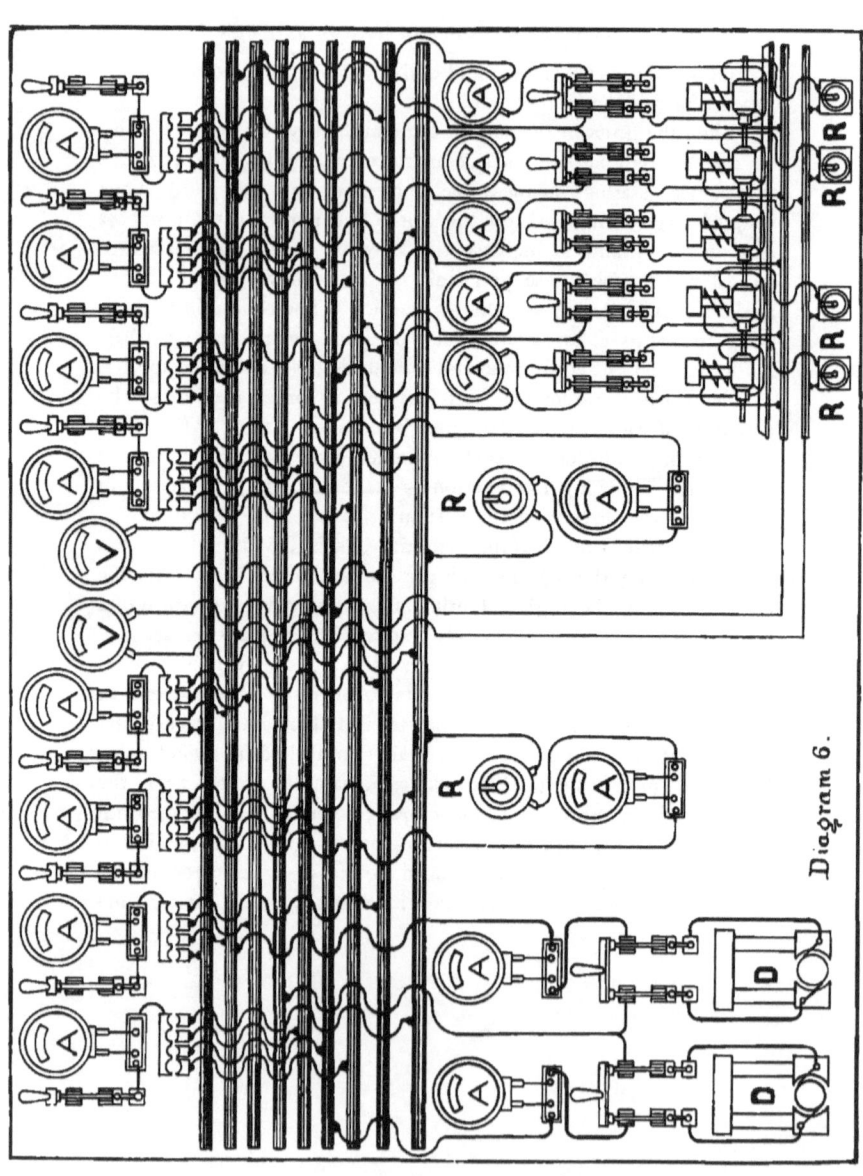

Diagram 6.

the potential of the current passing through it can be regulated by means of a field rheostat. Shunt wound boosters are usually used in low potential work which is controlled by the shunt field only.

In the three-wire system two boosters are required for each potential, and it is the usual practice to couple these boosters together, forming one continuous line of shafting, to which is also coupled a motor for the purpose of keeping the boosters up to full speed, and making good the losses due to transformation and the unequal loading of the high and low busses.

Diagram 6 shows the method employed for connecting boosters to a three-wire system.

It is evident that if one dynamo could be arranged to give several potentials, the efficiency of output would be raised and the losses inherent in the operation of the boosters would be saved. There have been several methods proposed for this purpose, but when a series wound armature rotates in a multipolar field, and a load falls on one circuit supplied by a section of this armature, a redistribution of magnetism will occur, caused by the unequal loading of this symmetrical armature; this will give an unequal potential delivery to the circuits from this armature, and a variation of potential beyond the control of the regulator. If this condition is avoided in the design of the machine, it becomes both expensive in first cost and uneconomical in operation; but with a symmetrical external distribution system, with regard to the station and feeders which are tapped radially into a concentric main, and if a load falls on one part of this main, and that section of the multipolar generator has its field increased, the distribution of magnetism in the concentric field will be identical with that of the distribution of current in the concentric main; in this way a compounding effect can be obtained external to the dynamo itself.

In certain cases, economy may be shown by inserting regulating resistances between the high bus, and busses of lower potentials may this way be obtained where the current demands on the busses of lower potential are small, and the losses in these resistances are less than that caused by a partly loaded engine and generator working on this reduced potential. This method also allows keeping the generator efficiency high by working all the units under the maximum possible loads for the different daily variable outputs of the station; this result can be obtained by the proper switchboard design, which will also effect a large operating economy.

In other methods of producing the several potentials required for close regulation on the external distribution system, the storage battery is one which is coming into use, and here the different busses obtain their potentials from different cells of the same series of storage batteries, and the potentials on these busses can be regulated by the battery-regulating switches. This gives the only method by which the regulation potential on the feeders can be affected without changing the loading on the generator.

As the feeder has to be supplied with several potentials, switching arrangements have to be provided so that these feeders can be connected at will to the various potential busses supplied by the generating apparatus. Where only two potentials are required, a double throw switch is used and the feeder is brought to the centre clip, when connection is made to either potential.

Where three potentials are used, double pole switches can also be used by grouping the nearby or low resistance feeders, so that they can be connected to common bus and low bus, and the outline or high resistance feeders, by means of a double throw switch, can be connected to common and high bus; the dynamos would, of course, have to have three-way switches in this case, so that they could be operated on any bus, as required.

Where the distribution system requires considerable variation in potential, a double throw switch could be used for a considerable number of potentials for the different feeders, if they are properly grouped, and if the drop is so slight at light loads that all feeders can be supplied from the common bus. Under these conditions all the busses can be tied together by tie switches, or all feeders can be thrown on one bus. Where the feeders themselves have to be supplied with more than two potentials during the load fluctuations on the station, several forms of switches have been devised for this purpose. The simplest form is a double pole double throw switch, having each blade on an independent handle; in this way a feeder can be supplied from any of four potentials, and also two busses can be connected together by this switch at low loads. See Fig. 83. The radial form of switch for the same purpose has different bus bars connected to clips and disposed concentrically around the centre; the switch plate is pivoted so that it can be thrown in on any of these busses when swung around in position. For this construction, see Fig. 84.

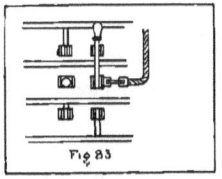
Fig. 83

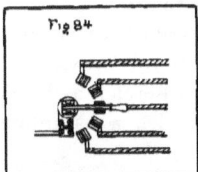
Fig. 84

Another form has been developed for this purpose, where the switch blade is detachable, and can be engaged with several terminals which align with the different bus terminals and the movable blade inserted; in this way the feeder can be connected to any bus. See Fig. 85.

In central station switchboards, there are several general methods of distributing the controlling and regulating of apparatus. One way is to concentrate all the regulators and field switches at one point, and the dynamo switches, ammeters and galvanometers on the dynamo switchboards.

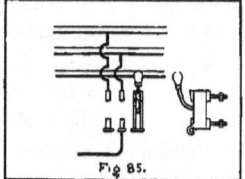

Fig 85.

The most prevailing method is to assemble the regulator, field switch, dynamo galvanometer, dynamo switch and ampèremeter on the dynamo board, all mounted on the same panel. The voltmeters are placed in some conspicuous position with the pressure switches; by using edgewise instruments, a density of four hundred kilowatts per running foot was obtained in the switchboard designed for The Chicago Edison Company by the author. The character of bus bar connections required for the carrying of one hundred and eighty thousand ampères from the lower dynamo board to the upper three-potential feeder board, which was located above the dynamo board in the Chicago Edison Company's station, is shown in Fig. 87.

The matter of field connections has to be carefully considered, for with large low potential units, a high induced potential is created when the field circuit is broken; this will tend to break down the insulation, and is the only way in which a hazardous potential can be produced in a low potential system.

The method of connecting the field in any generator will depend upon the system which is supplied by that unit. The self-exciting method is when the field is connected directly across the

ALBERT R. TURRO, E. E. CHICAGO EDISON COMPANY FIGURE 87

terminals of the generator; the rheostat is included in this circuit, and when the armature is rotated and creates a potential, due to the residual magnetism in the field, it sends a current circulating through the field windings, and in this way builds up the generator's potential. The bus exciting method is where the field is connected to the bus bars and excited by a potential external to its own. This method has the advantage of bringing the machine quickly up to the proper potential, and always of the right polarity; but when the field is to be withdrawn from the bus bar, after the machine is shut down, the field must first be short-circuited through a non-inductive resistance, such as a lamp bank, before it is broken, in order to reduce the potential of discharge.

The advantage of the self-exciting method is in the feature that, when the dynamo is shut down, the field dies away with the fall of potential on the generator; but in large multipolar units, the potential may rise very slowly, and for this reason bus exciting is largely used.

Separately exciting methods have no difference in their connection from the bus exciting, except a separate generator is used for field exciting. A condition arises where it is very advisable to have the field exciting of the generators so arranged that any of the generators of the station can be used temporarily as a separate exciter, when there occurs a very severe load or a short-circuit on the external distribution system. It is necessary to provide this method in order to hold up the potential of the generators working on a short-circuit, as the reaction of the armature under these conditions is so great that it kills the field, and the generator loses its potential when the field is excited, and the distribution is supplied from the same bus bar, and the station falls flat. Diagram 7 shows how by using a special field switch, all the fields can be excited from any generator. This method is only advised to be used in the case of emergency or short-circuit, and the proper connections can

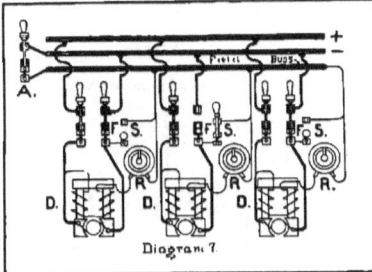

Diagram 7

be quickly made before the potentials have fallen, by inserting a dynamo on the field bus and withdrawing the switch that connects the field and main busses together.

Another method, known as the Donshea method, combines the good features of both the bus exciting and separately exciting methods. For connections see Diagram 8. The field switch "A" interlocks with one leg to the dynamo switch; to the other leg the field is permanently connected. If dynamo switch "B" and field switch "A" are thrown, the field of dynamo "B" is excited by the pressure on the busses. The dynamo is thrown on the system by closing dynamo switch "C." This switch is so arranged that when the main blade is withdrawn, it carries with it field switch "A," which is also electrically connected to it. When these interlocking switches are withdrawn, the field circuit is from the terminal of the dynamo, through field resistance box, field switch and dynamo switch

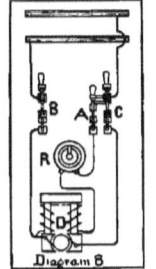

Diagram 8

blade, then back to the other brush of the dynamo. In this position of switches the dynamo is self-excited and the field will die away with the voltage on the armature.

The Potter method, Diagram 9, is applicable to compound generators, and by reason of the field being also excited by the series winding when thrown across the equalizer and positive side of the system, it will be in multiple with the series fields of the other dynamos in operation, and the current will be diverted to it, exciting the fields. By the aid of this initial excitation, the dynamo can be brought up readily to the proper potential by adjusting the shunt field, and when its proper potential is reached, it can be thrown in multiple with other generators.

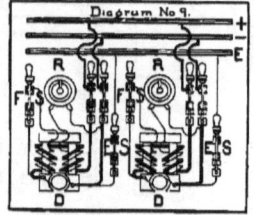

By practising this method of exciting dynamos, compound machines cannot be thrown in multiple without first connecting the series field, if the positive pole and equalizer be connected each to one leg, by double pole switches. For these connections, see Diagram 9.

RAILWAY SWITCHBOARDS

KNOWING the extreme conditions that arise in railway practice, it is required of the electrical engineer, in designing the switchboard for this service, to fully protect the generating apparatus from the shocks due to sudden overloads, and an automatic circuit breaker is in this case a necessity. The current passes through this circuit breaker, then through the ammeter shunt or the ammeter system itself to the negative bus bar.

In railway practice, the positive side of the machine, where the trolley is positive, is connected through the series winding. The equalizing connection is taken to the middle point of the switch to the equalizing bus, but the present practice in power stations is to equalize at the dynamo, and the equalizing switch either mounted on the frame of the dynamo itself or on a pedestal by the side of the dynamo. In other cases again, the equalizer is tied together permanently between all the dynamos. The disadvantage of having the equalizer opened is that there is a danger of the machine being thrown in circuit before it is equalized. In order to provide against this accident, several suggestions have been made; one is to make the switch at the dynamo double pole, through which are carried both the equalizer and positive connections; by means of this connection, the generator cannot be thrown in from the gallery or switchboard without having the equalizer thrown in first.

Another method has been proposed where the throttle of the engine is connected to the equalizer switch, so that when it is open it closes the equalizer switch; in this way the generator cannot be thrown in before it is equalized.

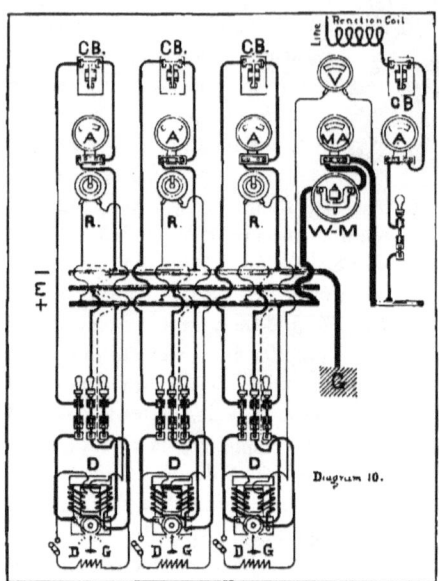

Diagram 10.

The field of the railway generator may be connected up in two ways; the one shown in full lines is the bus-excited method, and the one shown in dotted lines is the self-excited method. A dynamo galvanometer or voltmeter is arranged across the dynamo terminals of the dynamo switch, in order to show when the generator is of the right potential to be connected in on the system. Diagram 10 shows these connections.

It is also usual to allow for a panel between the dynamo and feeder panels, on which to mount the main ammeter, integrating wattmeter, voltmeters and pressure switches. The positive is only taken to the feeder board, and the feeders are provided with a single pole switch, ammeter, circuit breaker and reactance coil to choke back any lightning discharges and force the arrester to operate.

The dynamo panels should be provided with a small double pole lighting switch, where the station is lighted from the power generators, so that any generator can light the station independent of the power bus. This lighting circuit should be looped inside of the circuit breakers.

The present practice indicates that the best results obtained are when the lightning arresters are located as near the point where the feeders enter the station as possible. Behind the switchboard is not the proper place for the lightning arresters as a rule.

The panel form of construction is now universally adopted, the apparatus being mounted on an upper panel, with a foot plate about twenty inches high below it. These panels are made interchangeable for the different units and feeders, and the extension of a switchboard only requires that the bus bar and iron frame be extended. This gives a very flexible method, and amply provides for the future growth of the system.

Fig. 90 shows an assembly of a modern form of street railway switchboard. It consists of an edgewise dynamo regulator "A," dynamo switches "B"—the positive switch in this case being double throw, as this board is arranged for two potentials—field switch "C," dynamo galvanometers "D" and ampèremeters "E," also circuit breakers "F." This panel was designed by the author to take care of two two-hundred-kilowatt railway generators; its width is twenty-six inches, and all connections are made on the

RAILWAY SWITCHBOARDS

back of the board, as shown in the side and rear views. The equalizing in this case is done at the generator, and the copper bus bars are supported on the back by cast-iron brackets and insulated from them by marble blocks.

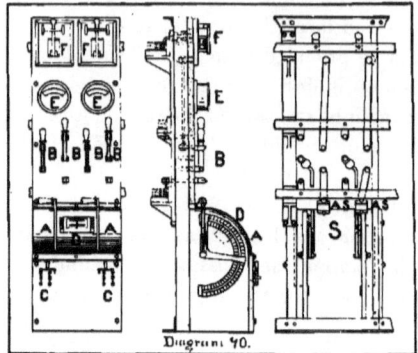

Diagram 70.

It is very useful in some cases to be able to separate the feeder systems, so that they can be supplied by independent generators, where extra demands of traffic require a higher potential to be obtained on the congested part of the system, in order that the schedule may be restored. To effect this result, the dynamo should be provided with a double throw switch, and also the equalizing system should be double, and the equalizer switch double throw. If the feeder switches are also provided double throw, the feeders can be operated on independent generators when required.

It is the usual practice to tie all rail and return grounds to a common negative bus; but to reduce electrolysis, in some cases, the ground returns which are tapped directly to the water or gas-pipe systems are brought to one ground bus, and the rail or return feeders are connected into a separate ground bus.

Generators are connected between the pipe return ground and the positive pole of the system, and the rail return ground to the positive pole of the system; these generators are maintained at such potentials, that the pipe systems are always kept at a lower negative potential than the rail system, so that all sneak currents will flow to and through the pipe system and be taken from the pipe system at the station where they can do no harm, for electrolysis takes place if they leave the pipe system and re-enter the earth in their attempt to return to the station.

In regard to the conductors behind the board, these are supported on porcelain insulators, or threaded through porcelain blocks as a rule, and in this case weather-proof insulation is sufficient for the conductor itself. All conductors should be stranded, and even the field wires should be a stranded conductor, in order to reduce the hazard that a solid conductor used here would increase. In some cases lead-covered leads are used, but where bare rubber is used for the insulation, great care should be exercised to prevent oil from reaching these conductors.

The thickness of the marble used for the switchboard surface should not be less than one and one-half inches where the circuit breaker opens on the board, as this blow will crack thinner marble. Care should also be taken not to drill too many holes in line, either horizontally or vertically, as it may seriously weaken the marble or slate through this line.

Exposed terminals of different potential, adjacent to one another, should be taped and insulated, or so shielded that no spark can jump between them, for when the circuit breaker opens on overloads there may be quite a rise in potential on the dynamo, which sometimes starts a flaring arc between exposed adjacent surfaces, which may produce damaging results.

So closely allied to the switchboard connections is the ground connection of lightning arresters, and so many good lightning arresters have been condemned on account of their poor ground connections, that a word here in regard to this important point will not be amiss. Every obstruction offered to the flow of this discharge by the ground wire subjects the station apparatus to an electrostatic stress, tending to break it down at its weakest point, and every means should be used so that the lightning discharge can jump the spark gap of the arrester and pass to ground. With this high frequency which a discharge possesses, it has a tendency to travel on the surface, rather than on the interior of the wire, and in this way choke its own passage; this effect increases as the diameter of the wire increases, and consequently only a small wire is used for the ground conductor, No. 6 being the usual size. A bend in a conductor greatly increases its self-induction, consequently the wire should be as straight as possible from the point of connection at the lightning arrester to the ground connections. Carrying this wire parallel to or near masses of iron will also tend to retard by self-induction the passage of this discharge to earth. To use a water-pipe system for earth is not the best practice; but where it is necessary, a brass lug can be clamped to the water-pipe and the contact surfaces amalgamated; into this lug solder the ground wire. After the connection is made, it should be painted over with two coats of air-drying asphalt varnish. No ground connections that are used for any other purpose should be used for the lightning arrester ground. No part of an iron structure or piping through the building should be used for the purpose of this conductor. The ground conductor should be connected to the water system, as near its entrance to the earth as possible.

A ground near running water or naturally moist earth will give the best results, but in all cases it must be below the frost-line.

If these cannot be secured, a hole can be sunk in the ground until water is reached. A copper plate two by two feet, with the conductor firmly soldered to it, will, in ordinary cases, be adequate for lightning ground. Loose waste metal does not materially increase the actual contact area of the earth plate. If such material is used for the earth plate, each piece should be connected with the ground conductor itself. The best material to use to get a low resistance ground is broken coke; this should be tamped well in the bottom of the hole to a depth of about two inches, and then the copper plate laid on this, and about four inches more coke laid over the ground plate and tamped well. Dirt can then be thrown over this and tamped lightly.

BACK VIEW OF
SPECIAL RAILWAY
SWITCHBOARD FOR
MARIETTA STREET
RAILWAY,
MARIETTA, O.
BUILT AND INSTALLED BY
WALKER CO.
CLEVELAND, O.

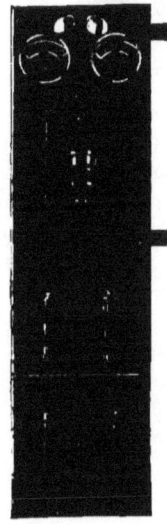

FRONT VIEW OF
SPECIAL RAILWAY
SWITCHBOARD FOR
MARIETTA STREET
RAILWAY,
MARIETTA, O.
BUILT AND INSTALLED BY
WALKER CO.
CLEVELAND, O.

ALTERNATING CURRENT SWITCHBOARDS

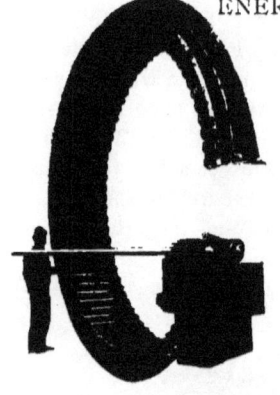ENERAL features in the construction of alternating current switchboards, wherein they depart in their designs from those already described, will be considered.

Increasing the potential brings in a hazard to the attendant which must be taken care of, also the leakage between terminals in a board designed to control over one thousand volts, if the marble itself is only depended upon for insulation. All terminals and screws or bolts holding these terminals to the marble should be insulated from the marble itself by mica or micanite, or an equivalent insulation; fibre is not of any value here, as it is a poor insulator when under compression. This method of insulating will greatly reduce the surface leakage, which may become serious in damp weather.

All exposed terminals on the front of the board should be screened from the switchboard attendant. Several fatal accidents have occurred from the neglect of this point, as the attendant has fallen against the exposed terminals with serious results. If these points be borne in mind, the alternating current switchboard becomes very simple in design.

The severing of high potential circuits, when alternating, is not so difficult, as the current reversals do not maintain such a fierce arc between these switch terminals, as in the case of much lower direct potentials.

In the switches used, it is usual to place over them a marble slab, having the free switch blade in front, and with recesses in this marble which screen the clips. Another form of construction is to have only the switch handle on the front of the board, and all the switch mechanism arranged behind the board, with a handle projecting in front which can be pushed or pulled to shift the connections from one side of the system to the other.

Alternating current machines, not producing a character of current which will magnetize their own field, require separate exciters and a separate field system. Also as alternating current dynamos, they are not, as a rule, run in parallel; this again separates the feeder system, so that each feeder can be supplied by any generator. The character of the current also allows of regulating devices which increase or diminish the potentials supplied that feeder, by inserting in it transforming, regulating, or compensating devices.

In long-distance transmission, where high potential is necessary in order to reduce the copper line costs, the alternators deliver the current at normal potentials to step-up transformers, which again increase the potential and reduce the current for a given kilowatt output.

All these characteristics and peculiarities of alternating current systems have to be taken care of in the switchboard design, besides the different connections required by the two-phase, three-phase, polyphase and monocyclic. First taking up the field connections and exciting methods where there are several alternators, it is better economy to use one exciter of sufficient capacity for all the alternators than a separate exciter for each machine; providing two exciters of sufficient capacity gives a duplicate exciting system.

A rheostat is used in the exciter fields, and all the alternators on this exciter can have their potential raised or lowered together. Each alternator is again provided with a rheostat in its field circuit, so that each alternator can be independently regulated. Where the alternators are not worked in multiple, a pair of horizontal bus bars are used for each single phase machine. The feeder is provided with a double throw switch, the middle clip of which has the feeder connected to it. The other terminals are provided with a plug receptacle, and also the bus from the generators. In this way the feeder can be connected to any bus, or changed to any other dynamo by first plugging in to the proper generator and then throwing over the switch of that feeder. Where the alternators are worked in parallel, only a double pole switch is required for the feeders, as alternators are added as the load of the feeders increases.

In order to accommodate the different combination of phases, the proper switching arrangements are shown in the diagrams for these different systems.

In working alternators in parallel, it is necessary to have both the potential and the period at which it occurs in step, in order that the generators may be thrown together and work synchronously. Two generators considerably out of step will jump together when connected in multiple, but will throw considerable strain on the armature and driving mechanism. It is also evident that a potential indicator will not be useful in putting alternators together, consequently a synchronizer is necessary. Fig. 91 shows a form of visual synchronizer, in which there are two primaries, one actuated by the electro-motive force of one generator, and the other by the electro-motive force of the other generator; each pri-

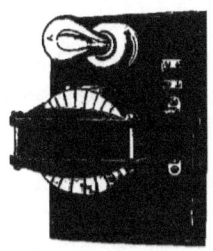

FIG. 91

mary induces an electro-motive force of fifty volts. When these two primaries are acting on the same secondary, and both currents are in phase, the electro-motive force will be one hundred volts, and the lamp will be maintained at full candle-power. The connections usually employed are those to reverse one of these primaries, so that the induced electro-motive forces oppose each other, and the lamps are out when the generators are in phase. The objection to this connection is, that if the lamp happens to break or the circuit be open, the alternators will be thrown in when they are probably out of step.

An acoustic synchronizer is also used, in which the currents from the two machines to be thrown together act oppositely on a diaphragm. When there is a phase difference between the two currents, the diaphragm is in vibration, but when the phase on the two machines is synchronous, the acoustic synchronizer does not emit a sound, as the attractions are equal and opposed. Fig. 92 shows this form of synchronizer.

FIG. 92

The phase indicator is an instrument that shows the angular difference which occurs between the maximums of two varying currents. Fig. 93 shows an instrument for this purpose, in which the current from the two sources, when they are out of phase with each other, tend to rotate an armature; this effort is proportional to the phase difference.

Voltmeters for alternating currents do not, as a rule, measure directly the potentials of the circuits, but a transformer is used with a known ratio of transformation, which reduces the potential of the circuit to fifty or one hundred volts. A voltmeter is connected across the secondary of the transformer, and the machines regulated by this voltmeter.

FIG. 93

Alternating ground detectors act in the same manner when one end of the primary is connected to the line and the other end to ground; if there is any leak on the system, a current will flow, and for the brilliancy of the lamp, the amount of external resistance to ground can be judged.

In regard to conductors used for alternating currents, ones having diameter larger than one-half inch should be stranded, as beyond this size there is considerable more drop than that due to the ohmic resistance of a conductor. Owing to the tendency of the alternating currents to distribute themselves unequally across a section of a conductor, and flowing more on the external surface rather than the interior of the conductor, this effect decreases as the frequency is reduced. Feeders were formerly protected by fuses, which were enclosed in a fuse box having a semi-circular groove in which the fuse was laid.

ALTERNATING CURRENT SWITCHBOARDS

Circuit breakers are now being used extensively on the alternators, and afford them the same protection which is so necessary in railway practice.

The connections for the most prevalent systems used in practical work are shown.

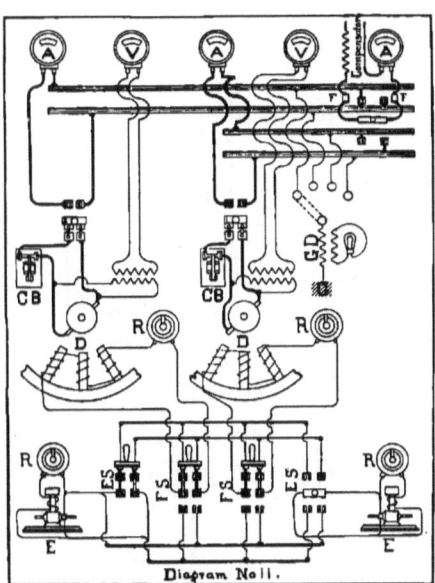
Diagram No 11.

Diagram No. 11 illustrates the method of connecting a single phase system having a common field circuit and independent dynamo circuits. Tracing out the field connections, the current generated at the exciter "E" passes to the terminals of the switches "S," "ES," and when this double throw switch makes connection with the field busses, the current is returned back to the exciter through the other pole of the switches. In this way the fields can be operated on either exciter and shifted from one to the other without opening the field circuits. These connections are entirely independent of the armature and its potentials, one-hundred-and-ten-volt system being

ALTERNATING CURRENT SWITCHBOARDS

generally used for exciting; the construction of this part of the alternating current switchboard can be followed out on the lines given for low potential switchboard construction.

Regarding the generator terminals, one is connected to the circuit breaker, then to one pole of the double pole switch, and carried through the ammeter to that alternator's bus; the other terminal of the alternator is also carried to the double pole switch to that alternator's bus. In this system, each machine is provided with a pair of busses, extending preferably behind the board, so that the feeders can be connected to these different busses.

Where only two alternators are used, it is evident that double throw double pole switches will make the proper connections; but when more alternators are used, it is the usual practice to provide plug terminals on the horizontal dynamo busses for each feeder, and also on the double throw feeder switch, so that this feeder can be plugged into any dynamo; by having this feeder double throw, in order to change from one dynamo to another, the idle terminals of the switch can be plugged into the dynamo to which this feeder is to be transferred, and in this way it can be shifted quickly without dipping the voltage on the light circuits.

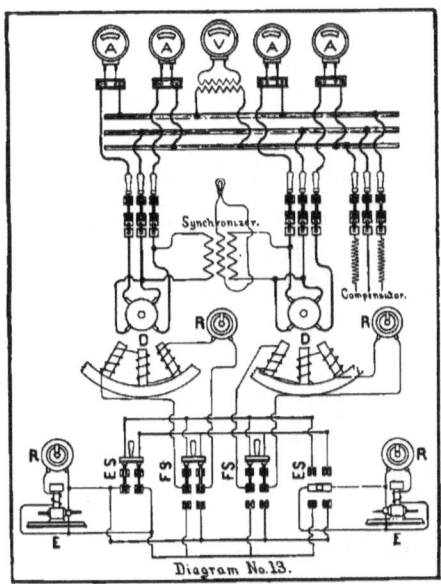

Diagram No. 13.

Voltmeters are generally provided for each alternator, and an ammeter for each feeder; also, where the external distribution requires, a compensator is introduced into the feeder circuit. By varying the inductance of the compensator, the voltage on the external system can be varied; a ground detector is also placed on this board with a multipoint switch, so that any circuit can be tested for grounds.

Diagram No. 12 shows the connections used where the alternators are run in parallel; in this case the field connections are the same as those shown in Diagram No. 11. A double pole switch

ALTERNATING CURRENT SWITCHBOARDS

can be used for the alternators, and all the alternators connected to a single pair of bus bars; the feeders only require a double pole switch, and in order to throw the alternators together a synchronizer has to be added and connected between the alternators.

Lightning arresters are not shown in these diagrams, as they are usually connected to the feeder at its entrance to the station.

Diagram No. 13 shows the connections required by the two-phase three-wire system, with the alternators operating in parallel, and Diagram No. 14 shows the connections for a two-phase four-wire system. The connections for the two-phase three-wire system are the same as those used for the monocyclic and three-phase systems, the middle wire only being used for power circuits in the monocyclic system, and the two outside wires for lighting systems.

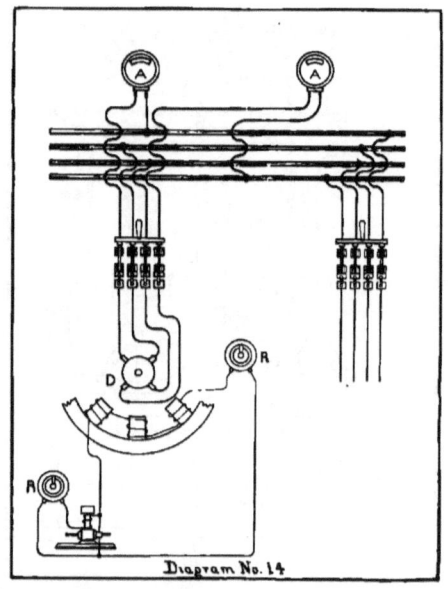

Diagram No. 14

ARC LIGHT AND SPECIAL SWITCHBOARDS

ISTORICALLY the arc light switchboard was the first combination of apparatus for the distribution of current for illuminating purposes which could be strictly called a switchboard. The requirements to-day have altered very little the constructional features from those originally installed, with the exception that the number of lights carried by a single dynamo has steadily risen as the art of insulation became more perfected, and to-day a one-hundred-and-fifty lighter is in practical operation, which means a potential of approximately seven thousand five hundred volts.

The same care in insulating the terminals from the switchboard itself has to be exercised, but the current quantity is low and does not produce a very serious arc when the current is broken.

The arrangement of the switchboard must be so flexible that any dynamo can be connected to any circuit, and any circuits can be looped together on any dynamo. The general arrangement to effect this is to bring all dynamo terminals to a series of plug receptacles which align with all the lamp circuit terminals. There are also transfer busses provided, as shown in Diagram 15. Here No. 1 machine is connected to No. 1 circuit; No. 2 machine operates No. 2 and No. 3 circuits in series; No. 3 machine plugs on the transfer bus, and circuits Nos. 4, 5 and 6 are in series, and

ARC LIGHT AND SPECIAL SWITCHBOARDS

the end of circuit No. 6 is transferred back to machine No. 3 by means of transfer bus. This shows the general combinations required on a switchboard. A number of forms of plugs have been devised for connecting these circuits. Fig. 94 is the oldest type, which is simply a plug with the dynamo or lamp circuit cable directly attached.

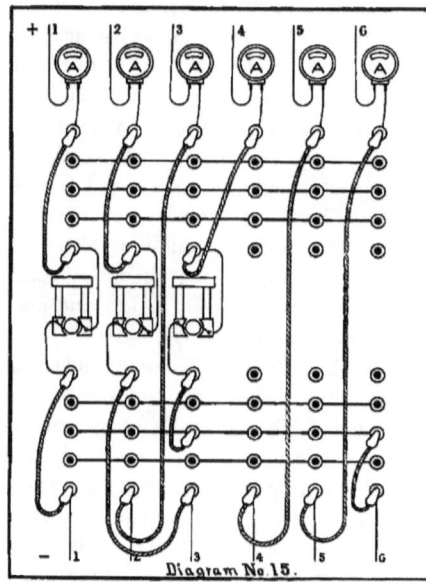

Fig. 95 shows a lock form, in which the plug is inserted between the spring and a notched latch, for the purpose of connecting these circuits together.

Fig. 96 is a plug only, without any connecting cable to it; the dynamo terminals are located on one face, and the lamp terminals on another opposite face about three inches apart. The plug is long enough to connect these two terminals, it passing through a spring bushing on the front face and registering with a tapering recess in the back face.

ARC LIGHT AND SPECIAL SWITCHBOARDS

Fig. 97 shows the form of plug, which, besides having the plug contact, also has a sleeve, which, when the plug is withdrawn, slides over the exposed plug terminal and screens this contact from the arcing effect.

The lamp circuit is provided with an ammeter, which generally has in connection with it a polarity indicator, so that the attendant can see whether he has plugged in his circuits correctly. Each leg of the circuit should be protected with lightning arresters.

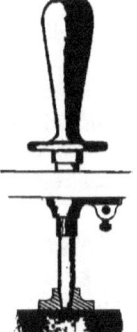

FIG. 94

The dynamo controller is generally located on a pedestal near the machine it controls; but in small plants these controllers are also assembled on the switchboard face with the rest of the apparatus. The general method of making these switchboards to-day is to separate the two potentials entirely, the positive usually being on the upper side of the board with the positive transfer busses, and the negative on the lower side of the board. In mounting these terminals on the switchboard face, the

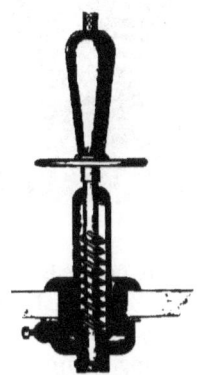

FIG. 95

FIG. 96

FIG. 97

same precautions should be exercised as are given for high poten-

131

ARC LIGHT AND SPECIAL SWITCHBOARDS

tial switchboards. A pressure switch is often used in connection with the arc light circuit, in order to determine the pressure across the terminals of the circuit, and also the number of lamps on the circuit.

By having a switch arranged so that one side of the voltmeter can be connected to ground, the insulation to ground can be tested while these circuits are in operation. Only the highest grade of insulated wire should ever be used to make these switchboard connections, and the flexible cables in front should be stranded of wires, not larger than twenty-two.

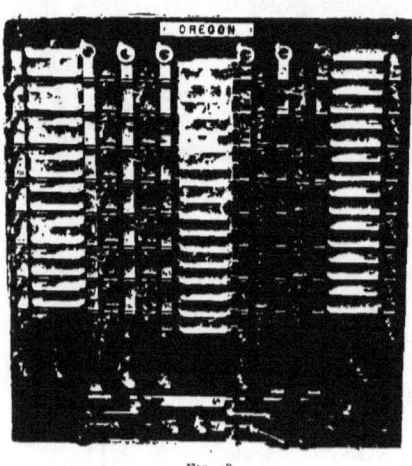

Fig. 98

There are a great number of special switchboards designed to fulfil special conditions, but they are not so generally used as to make their description of value. A few cases have been selected to illustrate this class where originality of design has been displayed. Fig. 98 shows the accepted form built for the Navy, where compactness was one of the essential features. It was also important that any circuit could be supplied from any generator independently.

The switchboard shown illustrates the arrangement for three generators and eight feeders; the knife switch employed here has a movable blade which can engage in the feeder terminal and be thrown so that any dynamo can be connected to any feeder. In this case the potentials are separated on two sides of the board—right and left—and the adjacent terminals are brought very closely together on account of them being the same potential. The dynamos are connected in multiple by the switches at the bottom of the cut, and the instruments and regulators are assembled on another board.

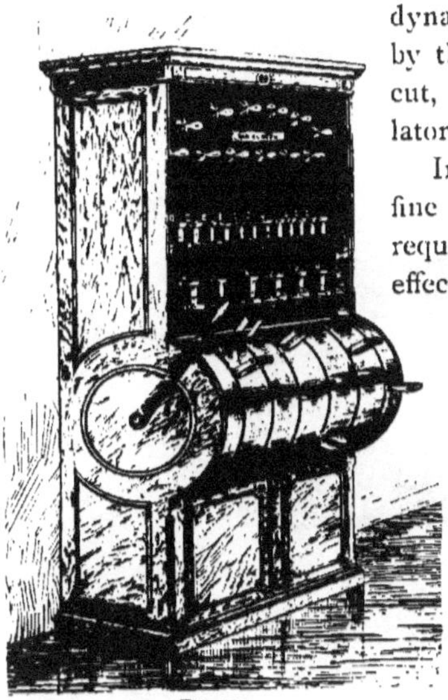

In the production of scenic effects, fine gradations of illumination are required, in order that the desired effect can be produced. This requires inserting a variable resistance in series with each of the lamp circuits to be controlled, and the regulators are interlocking, so that any group or groups of lamps can be varied together. Feeder switches and pilot lamps which are across the different circuits controlled, are assembled together for this special work.

FIG. 99

Fig. 99 shows one of the new types of switchboards designed for this purpose.

In alternating current work the resistance is substituted by a choking coil, which has a variable choking effect on the lamps in series with it.

ARC LIGHT AND SPECIAL SWITCHBOARDS

It has been the author's attempt in writing this essay on "Modern Switchboards" to give a practitioner's review of the art, and bring out such points in construction, appliances and connections as are necessary for their proper construction and design.

CIRCUIT BREAKERS
AND THEIR USE IN POWER TRANSMISSION

By W. H. TAPLEY

Chief Electrician U. S. Government Printing Office, Washington, D. C.

Reprinted from *The Electrical Engineer* by permission of Mr. Tapley and the Editors

When the application of individual electric motors to driving machinery became firmly established in the manufacturing world, and was conceded to be a more economical method of power transmission than belting with long lines of shafting, second only to the motor, and how properly to connect it to the machine which it was to operate, was the subject of suitable protection both to motor and machine.

The first thing that suggested itself, and naturally, was to protect the motor in the same way as lighting circuits, namely, to introduce a suitable fuse. This was done, and where motors were belted the results attending overloads were rather of an annoying and aggravating nature than anything which could really be called serious; yet when gearing and the direct application of armature to the main driving shaft of a machine began to supersede the belt, it was only a short time before the fact that a fuse was not an adequate protection became forcibly impressed upon the advocates of electrical power transmission.

As the art advanced, the thing to which the electrical engineer would turn for a rational solution of the problem was the electric current itself. How well this has been accomplished is shown by the successful introduction of the circuit breaker, now so universally used in all large power plants. That the magnetic property of the electric current was the means best adapted for the actuation of the protective device, and gravitation the most reliable force for governing its operation, is seen by the great superiority of the circuit breaker which depends entirely upon these forces, over those in which the effect of the actuating current is subject to variation due to extraneous conditions.

PROTECTION AND WHAT IT SHOULD BE, AS APPLIED TO A LARGE MANUFACTURING PLANT

In treating of this subject the tendency of the engineer has been to regard it almost entirely from an electrical point of view, incidentally, if at all, considering that which affected the real success of the manufacturing establishment employing motors, namely, constant service and lowest cost of production. Before entering further into this matter, let us see what is absolutely required to give a manufacturing establishment protection worthy of that name, when using electrical power transmission.

First—To secure the protection of electrical apparatus from motor to generator.

Second—To provide a method which will afford ample protection for the machinery to which electric motors are attached.

Third—To secure a freedom from interruption of production, and avoid the exasperating delay which is experienced in replacing any part of the protective device after the same has been called into service.

Fourth—After protecting everything in the shape of machinery, the safety of building and electrical apparatus and providing

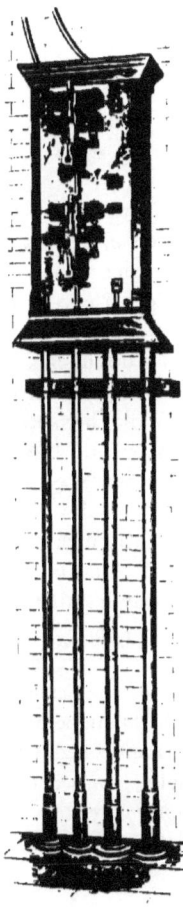

against the stoppage of production, the matter of reducing to the lowest possible point the liability of accident to the operators required to handle either motors or machinery must be considered, as indeed this is a matter of supreme importance.

That all the above-mentioned features are ever present, confronting the engineer, who is responsible for the successful operation of a manufacturing plant, is confirmed by the large number of so-called protective devices already offered to the public.

At present a very large part of the labor of the electrical engineer and the manufacturers of this kind of electrical apparatus has been directly in one line, that of protecting the electrical apparatus from the effects of overheating and the building from fires which might occur from heavily overloaded circuits.

As this feature is commanding a large share of attention in the electrical press and manufacturing world, it would seem best to devote our time in this article to the field suggested in the last three of the foregoing propositions, which, if satisfactorily solved, of necessity cover all the ground now under consideration by engineers, on the subject of proper and positive protection to electrical apparatus as applied to transmission of power.

CIRCUIT BREAKERS AND THEIR USE IN POWER TRANSMISSION

The pronounced success during the past two or three years of the direct application of electric motors to all kinds of machinery has put this method of power supply so far in advance of other methods that, notwithstanding the comparatively high first cost, it is now considered the most economical method, and should be adopted by every large manufacturing plant where the work is, in any sense, of an intermittent nature.

Let us now look from the electrical side to that of the manufacturing plant proper, and see if protection is not even more important and imperative here, where very much larger sums of money are invested, and until now have been wholly neglected except by insurance from fire.

To suggest something which may form a topic for discussion, let us take the case of a printing-press to which is directly connected an electric motor. The cost of the printing-press is, roughly speaking, $3,000.00, and that of the motor equipment $300.00. (These are nominal figures, which vary with the class of press used and character of work required from it; the cost of motors also varies considerably, but these figures are conservative and fully within the figures for which good apparatus can be purchased.) Allowing the electrical to be one-tenth the cost of the mechanical installation, does it not seem strange that it has always been the motor which it has been the sole idea of the engineer to protect, notwithstanding its cost is insignificant as compared with the value of the machine to which it is attached? Is it any wonder that the manufacturers of costly machinery, such as printing-presses, have looked with doubtful eye upon the method of direct motor application, whether it be by gearing or having the armature of the motor keyed to the main shaft of the press? The manufacturer well knew that for a short time the motor was capable of producing perhaps five times its rated output, and realized that if this period covered only a few seconds, there was a great probability that it would be sufficient to ruin the press should anything occur which would tend to stop it suddenly. He did not feel that there was even the protection which is afforded the presses when driven by belts, for these would slip when called upon to do much more than the normal work of driving the press.

The writer takes the same ground as the machinery builder, and when the representatives of companies manufacturing electrical apparatus were asked about this matter, they invariably assured the purchaser that a fuse inserted in the circuit supplying the motor with current would provide against all possible trouble of this kind. It was tried; the fuse worked in some cases and we began to take courage, thinking that perhaps we were too particular and that the fuse afforded the required protection. It was seen, however, that the blowing of the fuse might serve to protect the motor but not the press.

It is impossible to change over from one method of operating machinery to another without meeting failures, due perhaps to nervousness on the part of the operator when called upon to do a thing for the first time; and certainly it was so when motors were first used in this manner. The sudden turning on of the controller naturally blew the fuse, which too often was sufficiently increased in size to prevent this annoyance. But at its best the fuse served only to protect the motor, the requirements of the motor-driven machine being altogether overlooked. To better appreciate the shortcomings of the fuse in this respect, it is only necessary to understand the conditions under which it operates to open the circuit in which it may be placed.

The effective energy which must be supplied to the fuse is made up of the following quantities: First, heat sufficient to raise the temperature to that of the melting point of the fuse; and second, an additional amount of heat proportional to the mass and latent heat of fusion of the fuse, while in addition to this, the heat, being radiated by the fuse and its terminals, must be supplied. It will be seen from a consideration of these facts that the fuse requiring a relatively large excess of energy to effect its operation will permit a proportional excess of power to be supplied to the motor. The damage which may result from this may perhaps be more readily seen by an example. Suppose a foreign body gets into the working parts of a machine which is directly connected to a motor. The power supplied by the motor is now expended in the wrecking of the machine, the weakest parts yielding first to the strain. Only a very short time is necessary for the execution of great damage. The possibility of damage is then limited only by the energy required to blow the fuse.

It was only upon the advent of the modern circuit breaker that protection worthy the name was secured for motor-driven machinery. Owing to the much lessened energy required in the operation of this device, the time required, upon the occurrence of an abnormal flow, for the opening of the circuit is minimized, while, in addition to this, the heavier overloads are made to contribute some of their energy to the acceleration of the circuit-opening switch, thereby still further decreasing the time of opening.

It may thus be seen that by the use of a properly constructed circuit breaker the excess of power which may be communicated to the machine is vastly reduced as compared with the fuse. In fact, the time element is so lessened that the possibility of damage to machine as well as to motor is practically limited to that due to their combined momentum. What this is in each individual case makes it necessary to decide whether an auxiliary break to take care of the same is necessary or not. This is a question for the mechanical engineer to solve; but whatever this may be, it does not affect the principles set forth above, and only serves to bring out more clearly how necessary it is to shut off the current instantly and thus prevent the machine from acquiring any additional momentum.

As the question of machinery protection has been given so little consideration, it was deemed advisable to bring to the reader's notice, in rather a minute way, all the possibilities that it may have, in order to give to the subject the importance which we feel it deserves. In the foregoing, a reference was made to the comparative cost of a printing-press and that of the necessary electrical equipment to drive the same; *i. e.*, $3,000.00 for the former and $300.00 for the latter—a relative value of ten to one, which justifies the statement that the protection of machinery is a much more important consideration in electrical transmission than that of the motor. Can the electrical engineer afford to neglect this important feature of machinery protection and still hope that his customer will secure satisfactory results? for, after all, it is necessary that the new system, as a whole, shall be made more productive, and thereby more profitable, than the old.

It may be assumed that our third proposition is intended more especially for the manufacturer than the engineer; but, taking the ground that that which is of importance to the buyer concerns the seller also, we believe it is worthy of the close attention of both. As the writer has had an extended experience upon the application of electricity to printing machinery, he hopes to be able to treat this branch of a manufacturing business in a more positive manner than he could do should he endeavor to extend its scope into a more general or theoretical field.

Referring once again to the printing-press, with its electrical equipment cost (which we assume to be respectively $3,000.00 and $300.00), let us see what the production of such a press should be when it is running for three hundred days in the year on a fairly good class of printing, and what costly affairs stoppages of presses are, no matter what the cause. A press should earn an average of $10.00 a day, or $3,000.00 a year. This is not intended to express net earnings, but simply the average gross earnings for the press on commercial work, and we assume that it is running on such work continuously. It will be seen that delays caused by an accident to a press may prove much more expensive to the printer than the actual cost of the repairs to the press itself, as a delay of a week means a loss of $10.00 a day, or $60.00; and serious accidents often mean a month of working days, or $260.00, aside from the cost of repairs, which experience has shown are not to be lightly considered.

Nor is it to be forgotten that the press to which an accident usually occurs is generally running on a piece of work which must be completed within a given time. This means that we must lift the form and place it upon another press which has to be "made ready," perhaps interfering with other work, and all this additional cost must be borne by the manufacturer without any return for the same. Is not the manufacturer fully justified, then, in demanding that his machinery and output be equally considered with that of the electrical apparatus in the matter of protection?

If the protection of apparatus worth $300.00 is deemed so important as to occupy, as it undoubtedly does, the attention of the foremost electrical engineers, are we not justified in taking the position that protective devices should be so constructed as to fully protect the manufacturer at all points, and not stop with the electrical equipment alone?

By the use of the highest-grade circuit breaker now offered to the public, which fulfils in a very satisfactory manner all the requirements thus far considered, such a saving may be made, not only to the motor, but to the machine which it drives, that the loss occasioned by stoppages and on account of repairs will be practically eliminated, and the device will pay for itself many times in the first year.

The delays incident to the blowing and replacement of fuses are perhaps more annoying in newspaper and publishing offices (where mails have to be met and where the time for the completion of a particular piece of work is limited) than in most classes of manufacturing; but in any case the saving is so important as to amount to very much more than the cost of adequate apparatus. Indeed, it would seem to be the best paying insurance which the manufacturer could possibly obtain. Such delays as we have been considering are practically unknown when the fuse is replaced by a thoroughly mechanically and electrically constructed circuit breaker.

With the more universal adoption of the individual motor in electrical power transmission comes also the question of protection to operatives employed in handling the machinery. This is of so much importance that most of the States in this country have appointed inspectors to visit all manufacturing establishments and see that the proper precautions are used, and every means employed to lessen the danger to the employees, of whatever nature it may be. With all the precaution taken to prevent accidents, it is impossible to do away with them entirely. Strange as it may seem, the carelessness or foolhardiness of the employees themselves is mainly responsible for most of the accidents which occur to-day, thanks to the hearty co-operation of inspectors and employers in their endeavors to make accidents practically impossible. Yet I know of nothing to prevent a man from placing his hands in close proximity to running gears, and if he happens to have a piece of waste in his hands and the same gets caught, the resulting damage is only limited by the quickness with which the machine can be stopped. Several such cases have come under my personal observation, and only the prompt opening of the circuit breaker prevented very serious results. In one case a man's fingers were pulled into a train of gears, he endeavoring to clean the press while in motion. His fingers were badly jammed, but the opening of the circuit breaker stopped the press before any serious damage was done, and thus saved the man three fingers. Another case was that of a man who got his arm caught between the two cylinders of a press. He was fixing the packing on one of the cylinders and motioned to the feeder to reverse the press. Instead, she started it ahead suddenly with the result that his arm was drawn in between the revolving cylinders, and again, owing to the instantaneous opening of the circuit breaker the pressman escaped with a severely bruised arm, instead of crushed bones as we had all expected.

These serve as examples upon which to base a requirement that protection to operatives is not a matter of minor importance, and should not be put aside with the remark: "Let the employees keep their hands out of the machinery—we cannot protect everything and everybody." To be sure, we have no electrical or mechanical device which will make brains for the ignorant or prevent careless operatives from getting hurt; yet devices can be made, as we have seen, which reduce the results of such carelessness to a minimum, and their employment should become more general as the successful operation of them becomes better known.

Having thus considered in detail the protection demanded by a large manufacturing plant using individual motors as the method of power supply, we are confronted with the proposition : can these demands be met to the satisfaction of all concerned—to the builder of machinery, as well as of electrical apparatus, and to the manufacturer using them? If such is the case, what is the device necessary to fulfil the requirements, and is it commercially obtainable?

With reference to the fuse, we have only to read any of the scores of valuable papers written upon their action to fully justify us in writing against it "not satisfactory" and passing on. As the current itself may be the means of protection, as well as that of propulsion, the method of opening the circuit electrically should be employed. This has been successfully accomplished in the modern circuit breakers which operate on the inverse time element rather than the constant time limit.

The ultimate requirements of a circuit breaker are that we can rely upon it to do all that we have shown it should do, and operate successfully, not once, twice, or for a month, but always. When this ceases to be the case, the magnetic circuit breaker will be superseded by some other means of protection.

To produce such an instrument the highest skill, electrical and mechanical, is required. Long study of the existing state of the art and the conditions under which circuit breakers operate is necessary, and the closest attention must be given to every detail of manufacture.

With this an accomplished fact, such results are not the only reward, however, nor should they be. The public will cheerfully pay, not only for the labor and material used in its production, but also a profit sufficient to encourage the maker and enable him to continue the work, for the perfect is never obtainable and is only reached approximately. In electrical science, perhaps more than in any other, we are never able to write the word "Finis."

THE DEVELOPMENT OF THE CIRCUIT BREAKER

IT is an old saying that fire is a good servant but a poor master—a homely setting forth of the broad principle that Nature's forces become valuable to us only as we learn not merely how to harness them, but also how to properly restrain them in harness. In general, the more effective the force when under control the greater the danger to be apprehended should it overcome restraint. In pursuance of these principles, it is seen that the application of the physical forces to man's needs is invariably handicapped by man's inability to keep these forces to their proper channels.

An illustration of this is seen in the history of steam engineering. For many years after steam was first used as a motive power, its use was restricted to very low pressures, that it might be the more readily confined, and it was only with the application of the safety valve and improved methods of boiler and engine construction that the high pressures of later years became safe and practicable.

The domain of electricity furnishes us with another example. The one difficulty which must be overcome before the long-distance transmission of electricity becomes a complete success, lies in the absence of proper insulation, and it is only when this need shall be fully met that electrical transmission over long distances will become a prime factor in modern engineering.

These two instances are certainly ample warrant for the statement that the work of the engineer consists, not only in the devising of means by which Nature's forces may be directed into the

I-T-E CIRCUIT BREAKER
MERCURY CONTACTS, SINGLE POLE
5 TO 45 AMPERES

I-T-E CIRCUIT BREAKER
MERCURY CONTACTS, DOUBLE POLE
5 TO 45 AMPERES

service of man, but also in the adaptation of further means to the end that these forces may be readily restrained and kept from effecting damage.

As has been hinted, this fact is eminently true in the domain of electrical engineering. It is not surprising, therefore, that as the uses of electricity have vastly increased, and as the means for employing it have multiplied, there has been a marked development in the methods and devices employed for the protection of electric circuits.

In the pioneer days of electrical engineering, the introduction of strips of fusible metal into a circuit was supposed to afford that circuit ample protection in the event of an excessive flow of current. This device, however, was only good enough while there was nothing better to take its place. The objections to this method of protection are best appreciated by those most familiar with its results; for, aside from theoretical considerations, experience has taught that the melting of a fuse is dependent, not simply upon the passage of a certain volume of current, but also on the manner in which the fuse is connected into circuit, and the relation between it and the walls and cover of the fuse block. In very many cases the time consumed in the melting of the fuse, after the occurrence of a dangerous overload, was sufficient to allow great damage to be done in other portions of the circuit. Further than this, the replacing of fuses involves delays frequently inadmissible in this age of haste. These are but the more evident reasons why the fuse has given place upon the "Modern Switchboard" to the circuit breaker, an instrument which, in its approved form, involves no such uncertainties of action as are inherent in the fuse. Let it not be supposed, however, that the circuit breakers which were first introduced pos-

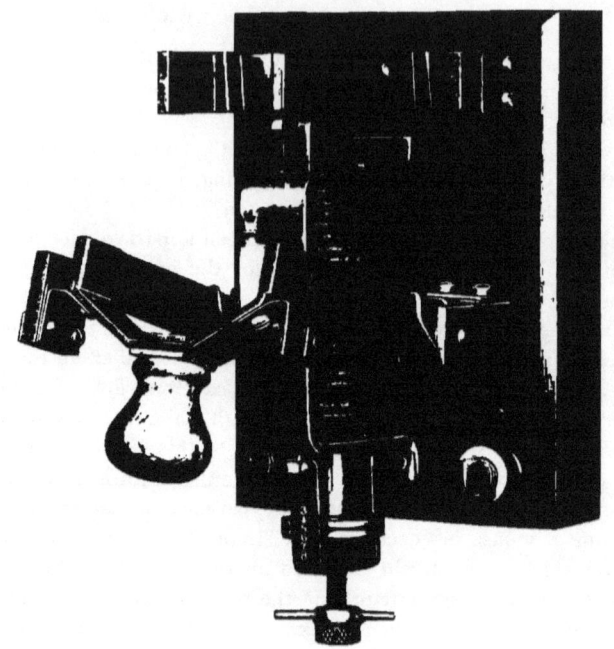

I-T-E CIRCUIT BREAKER
MIDGET JR. SINGLE POLE. 5 TO 25 AMPÈRES

sessed all the varied points of excellence which are realized in the perfected device; for while even the earliest forms offered a decided advance over former methods of protection, they still left room for improvement in many respects.

Those familiar with the first circuit breakers will remember that in their design the automatic features were secured at the expense of conductivity; in other words, a quality of primary importance for normal operation was sacrificed in order to provide for abnormal conditions.

Another, and no less vital weakness, evident not only in the first circuit breakers, but also in many later forms, was due to the fact that their operation was not entirely independent of everything save the actuating current. This was because such variables as the tension of springs and the friction between metallic surfaces were allowed to enter into or to influence the adjustment of these devices, and naturally resulted in much uncertainty in their operation.

One by one, however, these and other faults have been overcome, and the circuit breaker, in its most advanced form, combines with a conductivity not surpassed by that of the best switches of corresponding capacity, an accuracy of operation which compares favorably with that of the best ammeters, and a certainty of action comparable with that of gravitation itself.

Added to these developments there has been a vast increase in the scope of the circuit breaker. An instance of this is seen in the adaptation of the instrument to alternating current service. Until recently this has been considered to be out of the question, some of the reasons being that the earlier forms offered too great an impedance to alternating currents of even ordinary frequency, or heated unduly, owing to eddy currents or to hysteresis in the iron portions, while some of the instruments gave rise to a disagreeable humming noise when traversed by the alternating current.

However, after a careful study of the special conditions involved, these difficulties were traced to their sources and their remedies effected by adaptations of design and materials to the peculiar demands of the alternating current, resulting in the production of an instrument as perfect in its operation as the better known direct current circuit breaker.

Until recently, the operation of the circuit breaker has been limited to opening the circuit in the event of an overload; but in response to the demands of the ever-progressive engineer, there has been produced an instrument which effects the same end, upon the occurrence of a predetermined minimum flow or "underload." From this it was but a natural step to a combination of these functions in a single instrument, resulting in the production of a circuit breaker, operating upon the occurrence of either an underload or an overload, and this was accomplished without a sacrifice of any of the features entering into either the efficiency of operation or ease of manipulation, which characterize the simple overload instrument.

The single pole circuit breaker was a well-established success before a satisfactory double pole instrument was produced. The chief obstacle lay in the difficulty of obtaining a proper insulation between poles, without sacrificing the strength and the absolute rigidity necessary on account of the automatic action of the instrument.

Before persistent effort, however, these difficulties vanished, and the engineer has now at his command a double pole circuit breaker which fully meets the most severe requirements.

While these few cases indicate, in a measure, the lines along which the development of the circuit breaker has extended, there have been improvements of almost if not quite equal importance in directions other than those which have been cited. The circuit breaker has been adapted to service in circuits employing the highest pressures of ordinary practice, while in point of size it has been made to deal with the largest and smallest currents with equal efficiency and ease. A whole power station may be "opened up" by the operation of a single massive instrument, or the circuit of a single incandescent lamp may be automatically broken by a tiny device which may be covered by the hand.

To those familiar with the development of electrical apparatus, it is hardly necessary to mention that the continued progress of the circuit breaker has been effected only at the expense of time, thought and money, unstintingly applied. While it is no trifle in itself, the circuit breaker largely depends for its ultimate perfection on a multitude of details of seeming insignificance. Even such apparently unimportant matters as the depth of a contact, the exact temper of a switch blade, or the diameter of a bearing pin, are carefully thought over and planned, while the quality, quantity and disposition of the various metals entering into the electrical and magnetic portions of the instrument are made the objects of careful and exhaustive test. The designs of the circuit breaker are the outcome of long experience and close research, while the exactness of their construction and the beautiful finish, by virtue of which they enhance the appearance of even the handsomest of "Modern Switchboards," is the product of the most careful workmanship, aided by the most improved machinery that skill can devise or money secure. W. M. SCOTT, M. E.

INDEX TO ADVERTISERS

American Circular Loom Company,	181
American Electrician,	217
Baker & Co.,	199
Bibber-White Company,	189
Billany & Cochrane,	170
Blackwell, Robert W.,	183, 203, 221
Cutter Electrical and Manufacturing Company,	172, 173, 174, 177
Columbia Rubber Works Company,	195
Electric Porcelain and Manufacturing Company,	192
Electrical Engineer,	211
Electrical Review,	215
Eynon-Evans Manufacturing Company,	190, 191
Fairchild & Sumner,	217
General Incandescent Arc Light Company,	197
Hansell Spring Company,	171
Hartford Machine Screw Company,	183
Hill, W. S., Electric Company,	201
Hope Electric Appliance Company,	175
Imperial Brass Foundry, Limited,	218
Johnston, The W. J. Company,	213
Jones, J. & Son,	179
Kirkland, H. B.,	157
Machado & Roller,	187
Merchant & Co., Inc.,	219
Murdock, Wm. J. & Co.,	209
Moore, Alfred F.,	221
McLeod, Ward & Co.,	193, 217
New York Electric Equipment Company,	205
Partrick, Carter & Wilkins,	222
Phosphor-Bronze Smelting Company, Ltd.,	199
Porter & Remsen,	170
Roberts, H. C., Electric Supply Company,	185
Sibley & Pitman,	221
Solar Carbon and Manufacturing Company,	199
Stern, Edward & Co., Inc.,	163
Street Railway Journal,	195
Swoyer, A. P. Company,	193
Vallee Brothers & Co.,	203
Weston, Wm. H. & Co.,	167
Weston Electrical Instrument Company,	158, 159, 160, 161, 162
Wirt, Charles,	165
Zimdars & Hunt,	169
Zurn, O. F. Company,	199

MR. H. B. KIRKLAND

will be pleased to extend any information required as to I-T-E Circuit Breakers or C-S Flush Switches . . . . .

120 LIBERTY ST.,   NEW YORK

Representing
The Cutter Electrical and Mfg. Co.
American Circular Loom Co.

WESTON ❧ ❧ INSTRUMENTS

THE WESTON ELECTRICAL MEASURING INSTRUMENTS created a new epoch in the art of electrical measurement.

They were the FIRST, and remain the ONLY instruments which fulfil all the requirements of the Electrical Engineer, the Station Manager, and others engaged in the electrical business.

Since their first introduction to the electrical fraternity, ten years ago, their use has steadily extended to all parts of the civilized world, and they are now used and recognized as standards for all classes of work.

We have constantly endeavored to improve upon the original models, and have steadily made advances in methods of production, in details of construction and in electrical and mechanical design, until at present our instruments are vastly superior in all points to our earlier types.

We have also constantly added to the variety of styles and ranges, and have developed new forms suitable for all classes of work, until we now produce no less than one thousand different styles, ranges and models, and we are adding to our lines as rapidly as this can be done, considering the great labor and careful scientific investigation required to produce really trustworthy electrical measuring instruments.

We are original workers in this special branch, as is evidenced by Mr. Weston's discoveries of the negligible temperature coefficient alloys and standard elements; and all our work is of a class distinguished for its excellence of construction, adaptation to its special purposes, and originality of conception.

We consider the interests of your customers and desire to serve them faithfully, and our effort has been to build up a solid business to stand for all time, and doing this, we have put more money into original work, and in special tools and plant for the production of high-class electrical measuring instruments than any other concern in the world.

··· Weston ···
Electrical Instrument Co.

114 to 120 WILLIAM STREET
NEWARK, NEW JERSEY . . .

Weston
Instruments

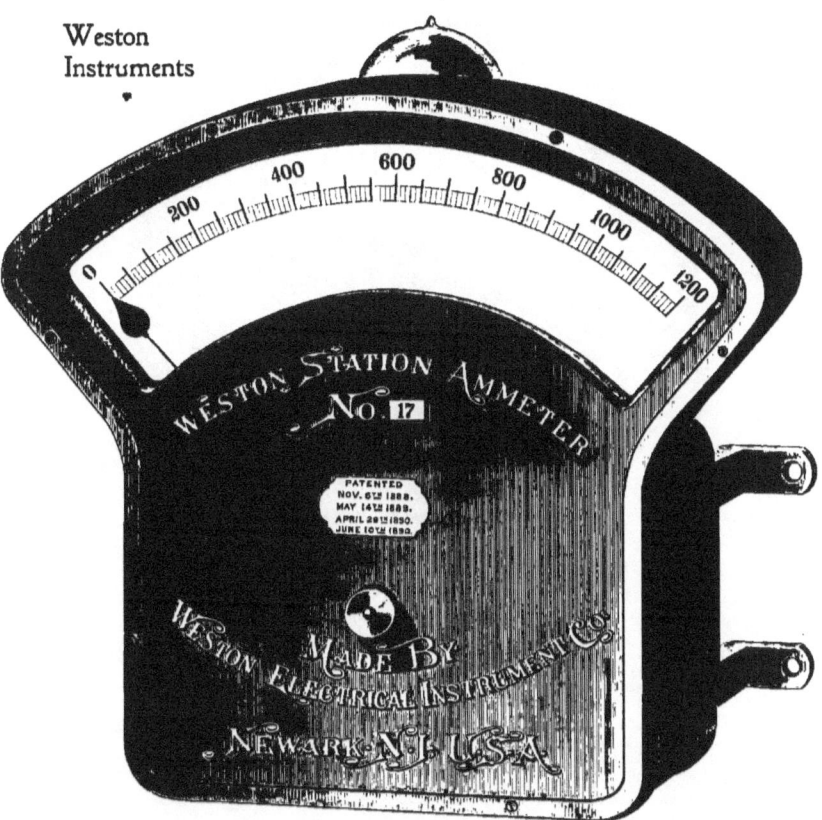

WESTON STANDARD ILLUMINATED DIAL STATION AMMETER, HALF SIZE

THE ILLUMINATED DIAL STATION AMMETERS AND VOLTMETERS are accurate, reliable and economical to operate.

The Ammeter is connected to a special alloy shunt, separate from the instrument, and placed at the back of the switchboard, or a short section of the mains may be used instead. Only very small wires are required to connect the instrument to the shunt. There is, therefore, a great saving in copper and labor in installing over other instruments, where it is necessary to carry the whole of the working current to and from them. These facts should be taken into consideration in connection with the price.

If at any time the capacity of the station should be increased beyond the capacity of the instrument, all that would be required would be to have it readjusted, since by its construction, the same instrument can be made suitable for any range.

The Voltmeters are very high resistance instruments, and are consequently extraordinarily economical of power.

These instruments have no "magnetic lag," are very "dead-beat," and are extremely sensitive and accurate. They can be left constantly in circuit with no material change in correctness.

The working parts are inclosed in an iron case, which effectively shields the instruments from disturbing influences of external magnetic fields.

WESTON ELECTRICAL INSTRUMENT CO.
114 WILLIAM STREET, NEWARK, NEW JERSEY

Weston Instruments

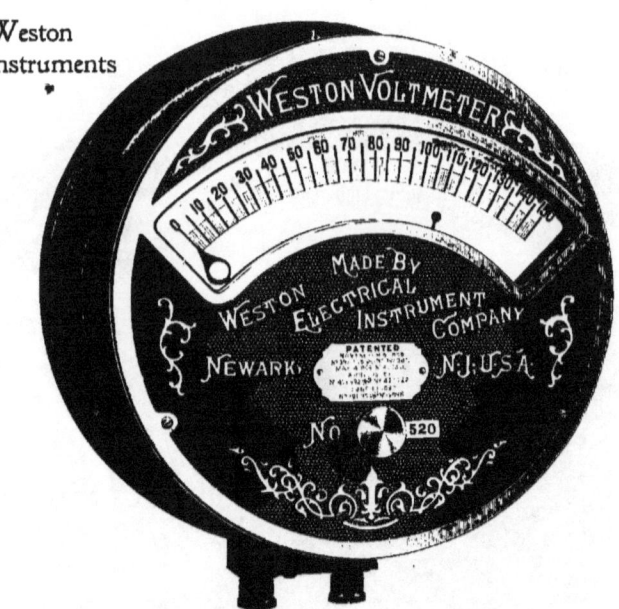

WESTON STATION VOLTMETER, "ROUND PATTERN," HALF SIZE

THESE instruments are identical in the principles of their construction with the Illuminated Dial Switchboard Instruments. The scales are shorter, and being drawn on opaque paper cannot be illuminated from the rear. They are the same in accuracy and reliability, and are also enclosed in iron cases.

For large plants, no type of switchboard construction possesses so many advantages as the

VAN VLECK EDGEWISE SYSTEM.

This saves enormously in FIRST cost of construction and erection, owing to the very small space required for the whole of the regulating, controlling and indicating devices. It also saves much labor in SUPERVISION and OPERATION.

It is the most convenient and easiest to manipulate, and closer regulation can be much easier maintained by its use than by any other system.

It facilitates the keeping of accurate output records, and reduces risks of error, as well as reduces costs.

We are sole licensees under the patents of Mr. Van Vleck and Mr. Weston, for the manufacture of all instruments for use with this system.

We strongly recommend the adoption of this system in all large plants, and it can be used in small installations with great advantage.

We make a full line of instruments adapted to the requirements of the system.

WESTON ELECTRICAL INSTRUMENT CO.
114 WILLIAM STREET, NEWARK, NEW JERSEY

Weston Instruments

In addition to instruments shown in the foregoing pages, we furnish the following for use on switchboards:

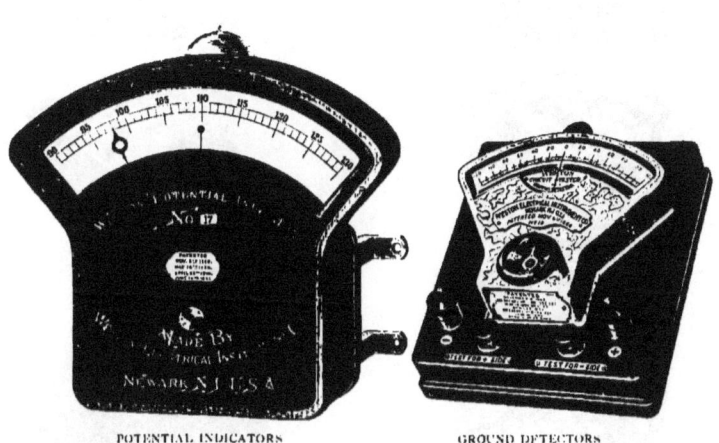

POTENTIAL INDICATORS   GROUND DETECTORS

WESTON ELECTRICAL INSTRUMENT CO.
114 WILLIAM STREET, NEWARK, NEW JERSEY

Weston Instruments

Our STANDARD PORTABLE INSTRUMENTS are all remarkably accurate, constant and reliable. They are all direct reading, are practically "dead-beat," and can be kept constantly in circuit without injury or change in accuracy.

Below we show illustrations of a few of the different styles of portable instruments we are manufacturing:

DIRECT CURRENT VOLTMETER

DIRECT CURRENT AMMETER

DIRECT CURRENT MILLI-VOLTMETER

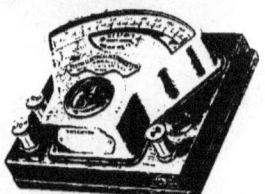

DIRECT CURRENT MILAMMETER

ALTERNATING AND DIRECT CURRENT VOLTMETER

WATTMETER

WESTON ELECTRICAL INSTRUMENT CO.
114 WILLIAM STREET, NEWARK, NEW JERSEY

Edward Stern & Co., Inc.

offer the services of a plant capable of producing all varieties of fine book printing, and the knowledge and experience which assure the highest accuracy and quality. Personal attention is given to the typography, binding, illustrations and the other important details of bookmaking. Estimates and suggestions will be furnished by mail or a representative upon application ✒ ✒ ✒ ✒ ✒

112-114 N. Twelfth St., Philadelphia

C. O. MAILLOUX
ELECTRICAL ENGINEER

METROPOLITAN LIFE INSURANCE BUILDING
NEW YORK

WESTERN ELECTRIC COMPANY
CONTRACTORS

The Wirt Dynamo Field Rheostat

REPORT OF TEST

LABORATORY OF QUEEN & CO., PHILADELPHIA, February 28, 1898

We append report on Wirt Rheostat, circular pattern, single disc, 12" diameter, Catalogue No. H22, catalogue rating 187 watts. We understood your instructions to test at five times the rated capacity in watts, with full resistance in circuit. Load to be thrown suddenly on cold rheostat.

	Start	30 min.	60 min.
Thermometer (bulb in outside contact merely),	70°F.	370	400
Watts on rheostat, ...	1080	1190	1230
Watts per square inch, ...	4.8	5.3	5.5
Overload (ratio to catalogue rating), ...	5.77	6.30	6.50

Insulation after test, hot, 1.6 megohms. After cooling to 212°, 5.4 megohms. Cold, 64.5 megohms.
General Condition after test, O. K.
Remarks :—Case of rheostat hot enough to scorch paper and to melt solder.

QUEEN & CO., Inc.
P. A. M.

Cast Iron Shell
Mica Insulation
German Silver or Nickel Alloy Resistance
No Joints in Resistance Conductor

Heavy Brass Switch Sectors
Phosphor-Bronze Contact Lever
Polished Bronze Wheel and Plate
Fifty Steps on Smaller Sizes
One Hundred on Larger Sizes

CHARLES WIRT
1028 Filbert St., Philadelphia

Send for description and prices

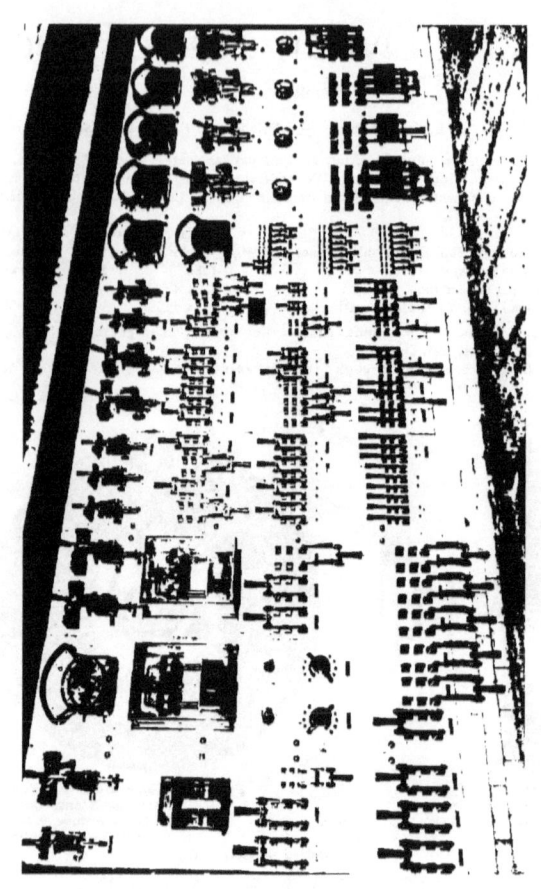

THE Switchboard illustrated on the adjoining page, installed in the engine room of Stephen Girard Building, Twelfth and Girard Streets, Philadelphia, controlling the electric light and power equipments of both the Stephen Girard Building and that of N. Snellenburg & Co., Twelfth and Market Streets, was designed and erected by

WM. H. WESTON & CO.

Electro-Machinists and Engineers

Keystone Spring Works Building

Thirteenth and Buttonwood Streets

Philadelphia, Pa.

TELEPHONE BUILDING
NEW YORK CITY

WESTERN ELECTRIC CO.
ENGINEERS AND CONTRACTORS

The Modern Switchboard Builders
...OF AMERICA...

ZIMDARS & HUNT, of New York

Desire to call attention to the fact that it is never necessary to specify as to quality when dealing with them, as it will continue to be their policy in the future, as it has been in the past, to manufacture only from the most approved designs, with the most skilled mechanics obtainable, and using only the best grades of materials the market affords.

SWITCHES *that are models of excellence, correct in design, accurately and carefully made, and embodying the results of an extended experience in switch manufacture.*

SWITCHBOARDS *that are too well known to require comment. The finest that skill can produce. They never fail to anticipate the specifications of the most progressive engineers.*

PANEL BOARDS *without an equal; easily in the lead for correct design, superior workmanship, and beauty of appearance; will show off to advantage anywhere; a credit to any installation, and* 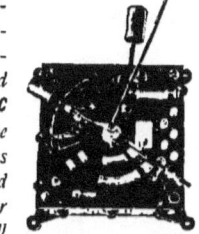 **AUTOMATIC SWITCHES AND AUTOMATIC MOTOR STARTERS**, *devices which have met with the most unqualified indorsement of all parties having any acquaintance with them. They are, beyond a doubt, the most approved articles of their kind ever offered, suitable for elevator, pump, crane, organ and all other work where automatic starting is desired. Made for direct or alternating circuits. Absolutely reliable in every respect.*

ZIMDARS & HUNT
MAKERS OF

High-Grade Electric Light and Power Specialties
127 FIFTH AVE., NEW YORK

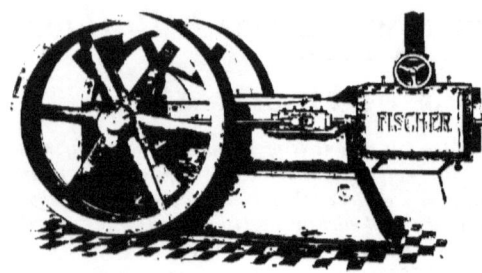

PORTER & REMSEN
ENGINEERS AND CONTRACTORS
39 Cortlandt St.
NEW YORK

Complete Steam-Power Equipment, High-Grade Engines for Electric and Manufacturing Plants, Nordberg Corliss Engines and High-Duty Pumping Machinery. Special Valve Gear for high speeds.

Gas and Air Compressors. Milwaukee Feed-Water Heaters.

Fischer Automatic High-Speed Engines, Single and "4"-Valve for Belted and Direct Connected work.

"Buffalo" Steam Pumps.

Nordberg Jet and Surface Condensers.

Nordberg Automatic Governors.

Walker's Metallic Piston Rod Packing.

INFORMATION AND ESTIMATES FURNISHED ON APPLICATION

Write us when in the market for Machinery

BILLANY & COCHRANE
DEALERS IN
Machinists' Tools, Light Iron and Wood Working Tools
BLACKSMITHS' AND JEWELERS' TOOLS

Belting, Packing, Hose, Bolts, Nuts, Washers and Screws of all Kinds

527 COMMERCE ST., PHILA.

TELEPHONE 1538

Hansell Spring Co.

NEWARK, N. J.

MANUFACTURERS OF ———

Railway and Machinery

Springs...

..OF..

High-Grade Crucible Cast Steel

Special Springs of Piano Wire
    for severe and unusual service

The C=S Automatic Switch

THE popularity of this, the only reliable automatic switch, is constantly increasing. It is specially adapted for DARK CLOSETS, TOILET ROOMS, etc. For this purpose it is set flush in the rabbet of a door-jamb in a manner similar to the well-known burglar-alarm spring. It is strictly automatic. Opening the door turns the light on, while closing the door turns the light off.

The Switch is also made with a "reverse action," i. e., opening the door turns the light off, while closing the door turns it on.

Another valuable application of the C-S Automatic Switch is shown here, wherein it is so arranged as to positively turn off the electric lights from a hotel guest-chamber every time a guest leaves his room and locks his door from the outside.

It is estimated that fully ninety per cent. of all hotel guests, upon leaving their rooms, invariably leave their lights turned on, and in a hotel of ordinary size only, this means a waste of thousands of dollars annually. This arrangement of the automatic absolutely prevents such waste, thus adding so much each year to business revenue.

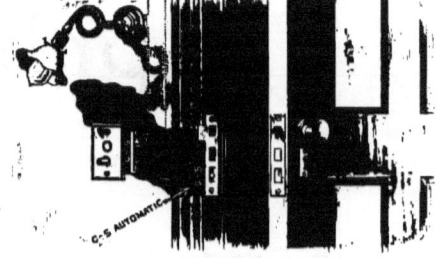

The plan of application and operation is simple in the extreme, as may be seen from the following

EXPLANATION.

All modern hotel guest-chambers are fitted with "secret" or double-bolt locks, one bolt being operated from the outside, the other from the inside of the door.

For this service of economy a C-S Automatic is placed in the moulding of the door-jamb, immediately back of the striker plate of the lock, in such manner that the throw of the bolt which locks the door from the outside opens the switch and turns off the light, while unlocking the door from the outside turns on the light. At the same time, locking the door from the inside has no effect upon the action of the switch. Arranged in this way, the automatic acts as the controlling switch for the room circuit, and locking the door from the outside will invariably turn off the lights.

For further information in regard to C-S Switches and Accessories, consult our catalogue.

The Cutter Electrical and Mfg. Co.
PHILADELPHIA AND NEW YORK

About March 1st
WE WILL BEGIN TO DELIVER ON ALL ORDERS OUR

NEW AND PERFECTED C-S SWITCH

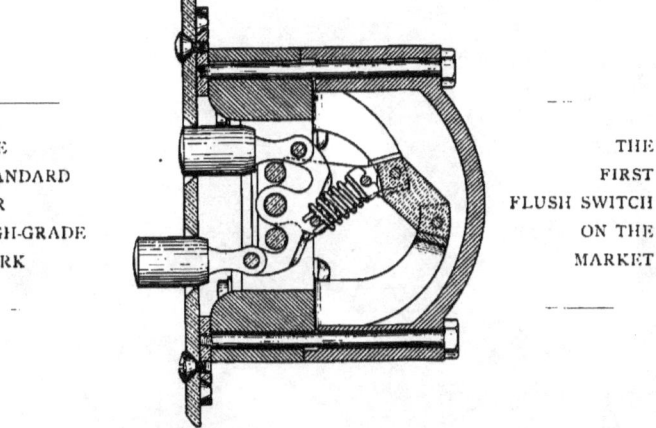

THE STANDARD FOR HIGH-GRADE WORK

THE FIRST FLUSH SWITCH ON THE MARKET

This Switch, as is invariably the case with an article with an established reputation, has served as a copy for numerous imitations, cheap and otherwise. Beware of them. They do not stand the test of time, and are very costly. RECENT CHANGES IN THE C-S SWITCH MAKE IT BETTER THAN EVER. THE **"PUSH"** HAS BEEN MADE **EASIER** AND THE FORM OF THE **SPRING** CHANGED SO THAT BREAKAGE OF THIS PART IS PRACTICALLY IMPOSSIBLE.

THE CUTTER ELECTRICAL AND MFG. CO.

New York, 120 Liberty St.   Philadelphia, 1112 Sansom St.

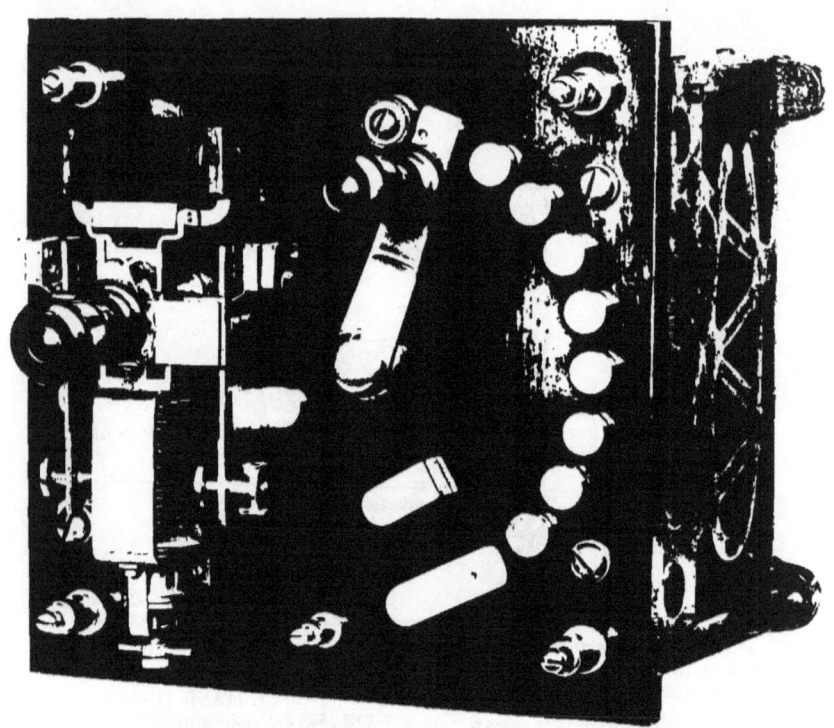

I-T-E MOTOR STARTER

A combination consisting of the usual means of closing circuit with gradually decreasing resistance, with a DOUBLE POLE AUTOMATIC CIRCUIT BREAKER connected therewith, the circuit breaker being specially designed to AUTOMATICALLY open the circuit upon a predetermined OVERLOAD or SHORT-CIRCUIT; also to OPEN CIRCUIT AUTOMATICALLY IN CASE THE CURRENT SUPPLY IS CUT OFF. AUTOMATIC means to PREVENT THE CLOSING OF CIRCUIT BREAKER UNLESS THE RESISTANCE IS ALL IN. The resistance-controlling arm to be operated MANUALLY, as opposed to the usual automatic methods employed; the spring part of this arm being used only to prevent the same from remaining on any of the intermediate resistance contacts. All contained in one device.

The Cutter Electrical and Mfg. Co., 1112 Sansom Street, Philadelphia

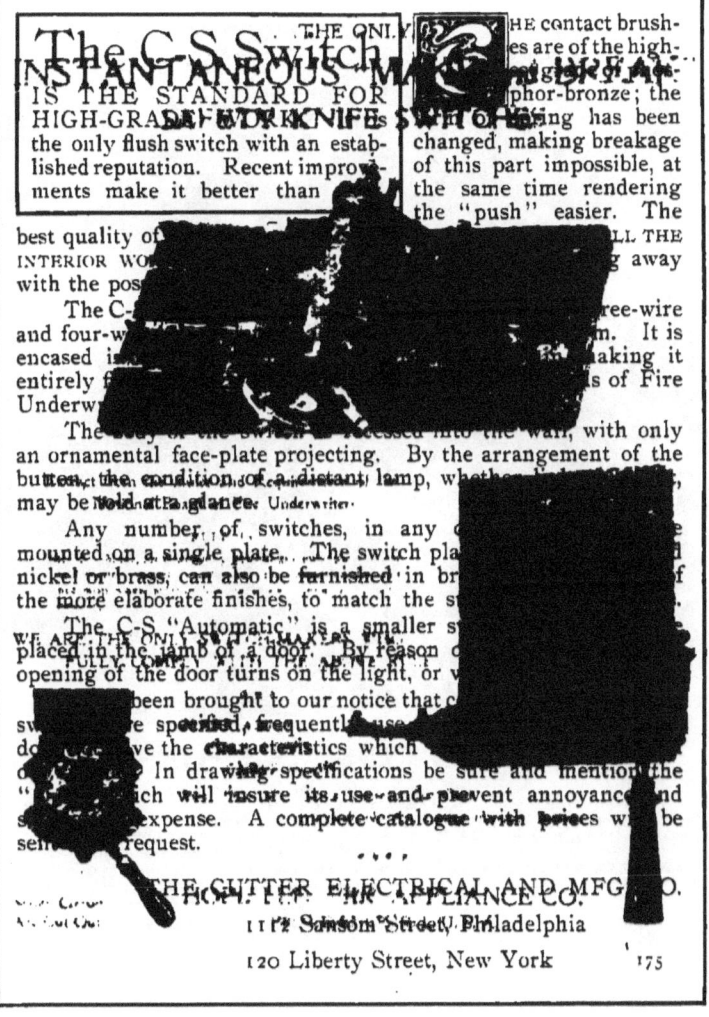

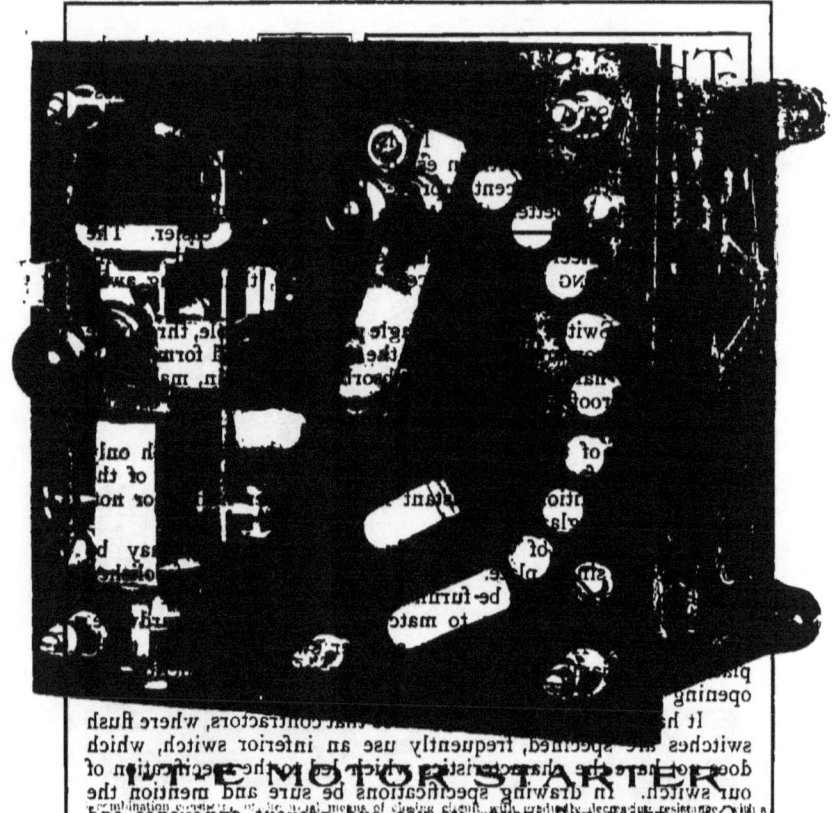

...THE ONLY...
INSTANTANEOUS "MAKE and BREAK"
SAFETY KNIFE SWITCHES

Extract from the Rules and Requirements of the
National Board of Fire Underwriters

RULE 43.

SEC. F. Must, for constant potential systems, have a firm and secure contact; must make and break readily, and not stop when motion has once been imparted by the handle.

WE ARE THE ONLY SWITCH-MAKERS WHO FULLY COMPLY WITH THE ABOVE RULE

✱ ✱ ✱

MAST ARMS
POLE STEPS
CABLE CLIPS
Four Pole Alternating Current Cut-Outs
Indestructible Arc Lamp Hanger Boards

✱ ✱ ✱ ✱

Series Circuit
Arc Cut-Out

HOPE ELECTRIC APPLIANCE CO.
PROVIDENCE, R. I., U. S. A.

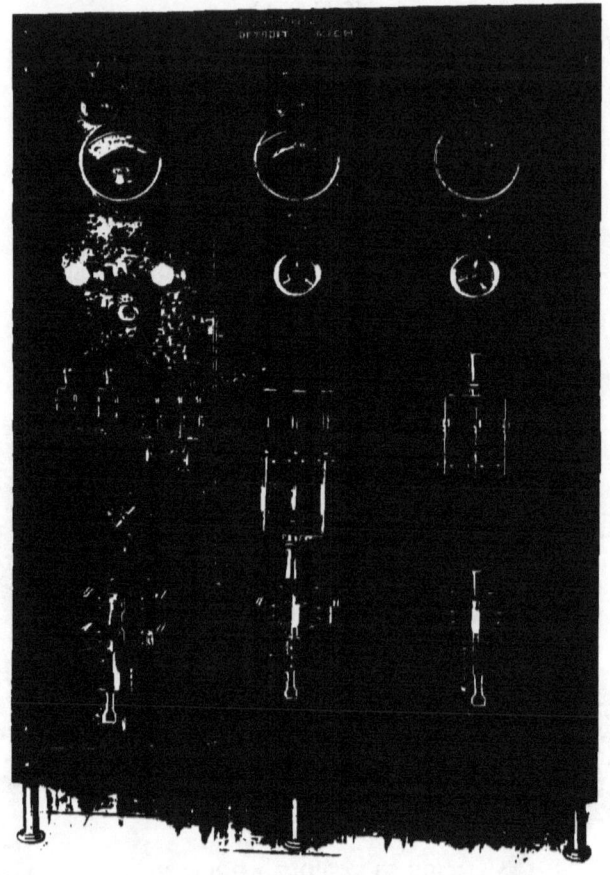

SMITH & CONANT
ELECTRICAL CONTRACTORS

HARPER HOSPITAL
DETROIT, MICH.

INSTALLED BY
MICHIGAN ELECTRIC CO.
DETROIT, MICH.

OLD COURT HOUSE
ROOM 7, FIRST FLOOR.

City of Boston.
Office of Electrical Construction Division
of Public Buildings Department.

Nov. 8th, 1897.

Cutter Elec. and Mfg. Co.

Gentlemen:

    It gives me great pleasure to notify you that the special double pole Circuit Breakers, which you built for the Boston City Hospital switchboard, have proved their value under a rather unexpected test. Last Wednesday evening one of our men accidentally dropped his wrench across the bus bars of a large tablet board, completely short-circuiting one side of the three-wire system, which is balanced by a 5 K. W. Motor Generator. The Circuit Breaker, which was set at 300 amperes, opened instantly, even before a 50 ampere fuse on the 5 K. W. unit could act. Had there been no Circuit Breaker, we should have lost our motor generator, and consequently the whole system. No damage resulted to the 5 K.W. unit, which had to deliver current sufficient to open the Circuit Breaker, and we suffered no interruption of service except on this one feeder.

    Your Circuit Breaker is ALL RIGHT.

                  Yours truly,

                  B. B. Hatch.
                          Engineer.

GEORGE H. PRIDE
ENGINEER AND BUILDER

EQUITABLE BUILDING
NEW YORK CITY

HIGH-GRADE SWITCHBOARDS PANEL BOARDS AND SWITCHES

SOME PLANTS, IN AND NEAR NEW YORK, FURNISHED WITH OUR APPARATUS

Proctor's Pleasure Palace, New York City
New York Orthopœdic Hospital, New York City
Dakota Apartments, New York City
Manhattan Electric Light Company, New York City
United Bank Building, New York City
National Meter Company, New York City
Hotel Waldorf, New York City
Staten Island Rapid Transit R. R. Station, St. George, S. I.
Midland Beach Casino, St. George, S. I.
Clarendon Hotel, Brooklyn
Pratt Institute, Brooklyn
F. Loeser & Co., Brooklyn
Crocker Wheeler Electrical Company, Ampère, N. J.
Hudson Electric Light and Power Co., Hoboken, N. J.
Jersey City Electric Light and Power Co., Jersey City, N. J.
Washington Light, Heat and Power Co., Washington, N. J.
Nepera Chemical Co., Nepera Park, N. Y.

J. JONES & SON,
67 CORTLANDT STREET,
NEW YORK CITY   FACTORY, BROOKLYN

MAJESTIC BUILDING
DETROIT, MICH.

CHARLES G. ARMSTRONG
ELECTRICAL ENGINEER

BUILT AND INSTALLED BY
MICHIGAN ELECTRIC CO.
DETROIT, MICH.

CLIFF HOUSE, SAN FRANCISCO, CAL.

MANUFACTURED BY

AMERICAN CIRCULAR LOOM CO.
CHELSEA, MASS.

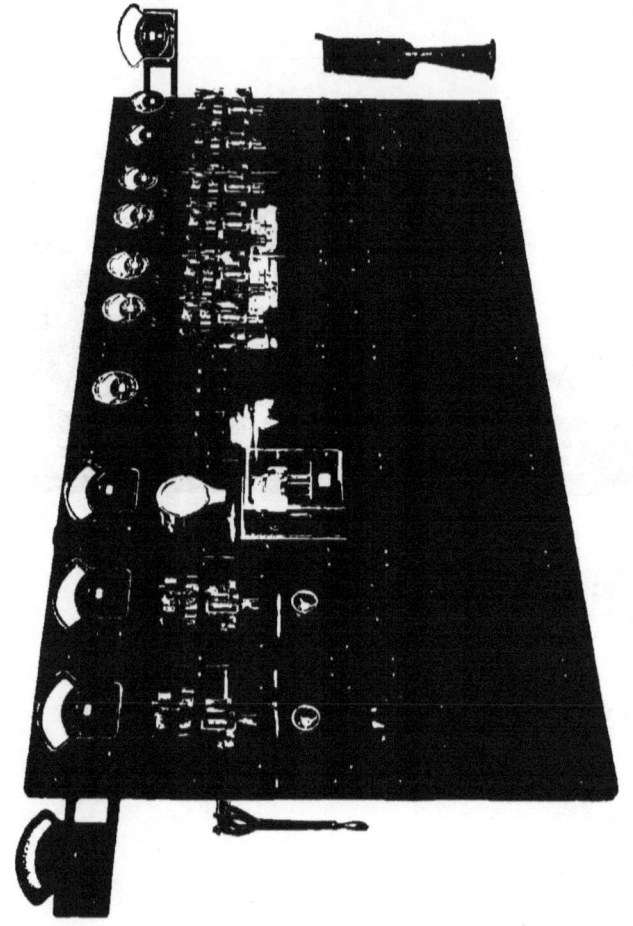

LORAIN & CLEVELAND RAILWAY CO.
FRONT VIEW

INSTALLED BY
SIEMENS & HALSKE ELECTRIC CO.
GRANT WORKS P. O., ILL.

ROBERT W. BLACKWELL

39 VICTORIA STREET, W.

LONDON, ENGLAND

Engineer and Contractor for Electric Tramway Construction and Equipment

Poles, Trolley Wire, Feeders, Rail-bonds, Insulators, Trolleys, Motor Trucks, Engines, Line Material and Supplies. I-T-E Circuit Breakers.

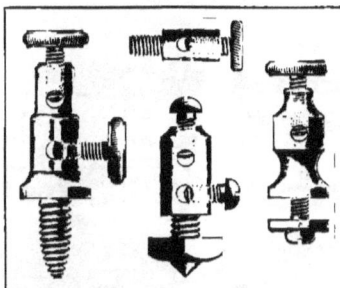

MANUFACTURERS of all classes of binding posts, with screws, nuts and washers for the same; also magnet cores and all other turned parts for electrical work not requiring stock more than 2½ inches in diameter. Hexagon, square and round head cap and set screws, from all kinds of material. All small turned parts for bicycles, guns, pistols, clocks, eye-glasses, watches, etc., etc. German silver, silver and gold screws made to order. We are headquarters for automatic machinery for producing all classes of turned work; also automatic and hand machines for finishing operations.

SEND FOR CATALOGUE AND PRICE LIST

HARTFORD MACHINE SCREW CO.
Hartford, Conn., U.S.A.

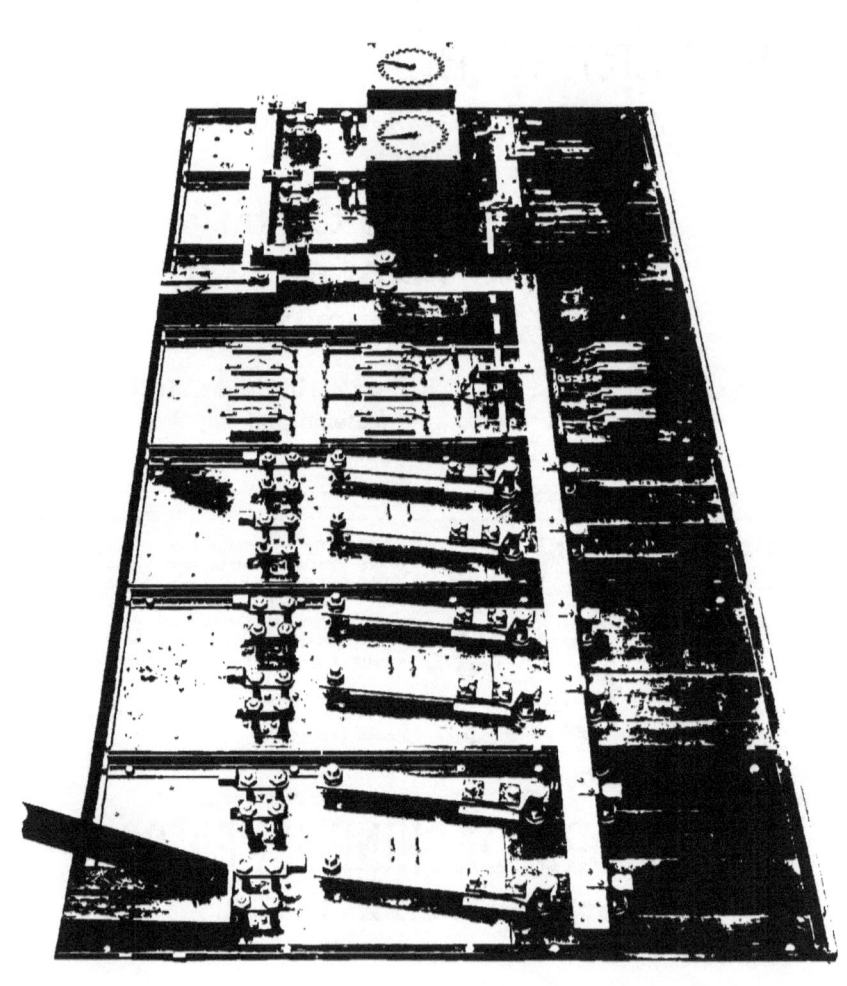

H. C. ROBERTS ELECTRIC SUPPLY CO.

·····The·····
Highest Class
Goods at the
Right Prices··

831 ARCH STREET
PHILADELPHIA

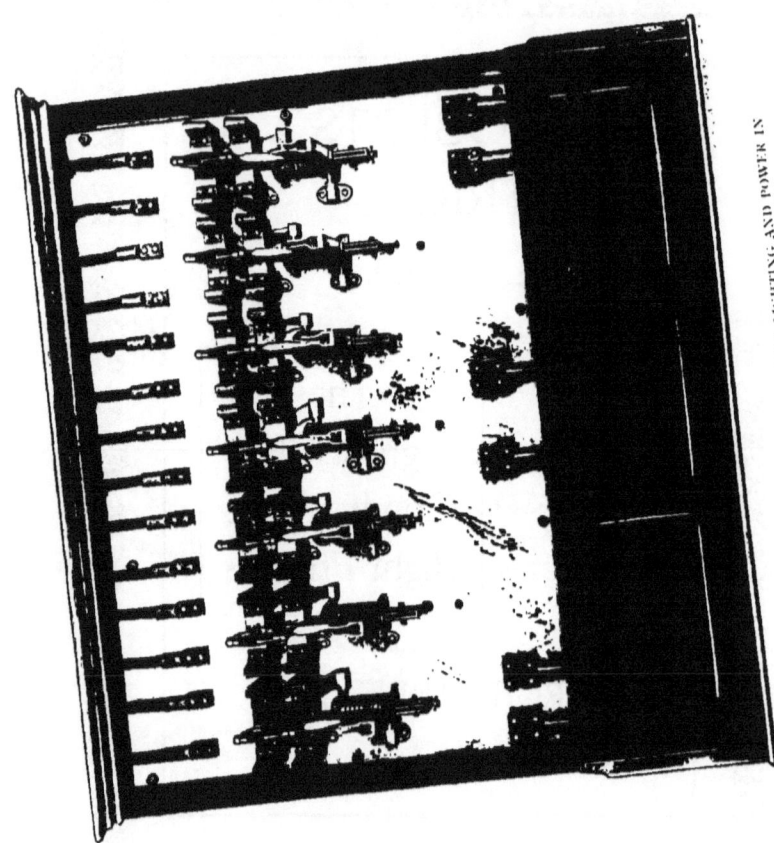

MAIN DISTRIBUTING CENTRE FOR BOTH LIGHTING AND POWER IN NEW WEST WING, GOVERNMENT PRINTING OFFICE, WASHINGTON, D.C. INSTALLED BY U.S. GOVERNMENT, NOVEMBER, 1896

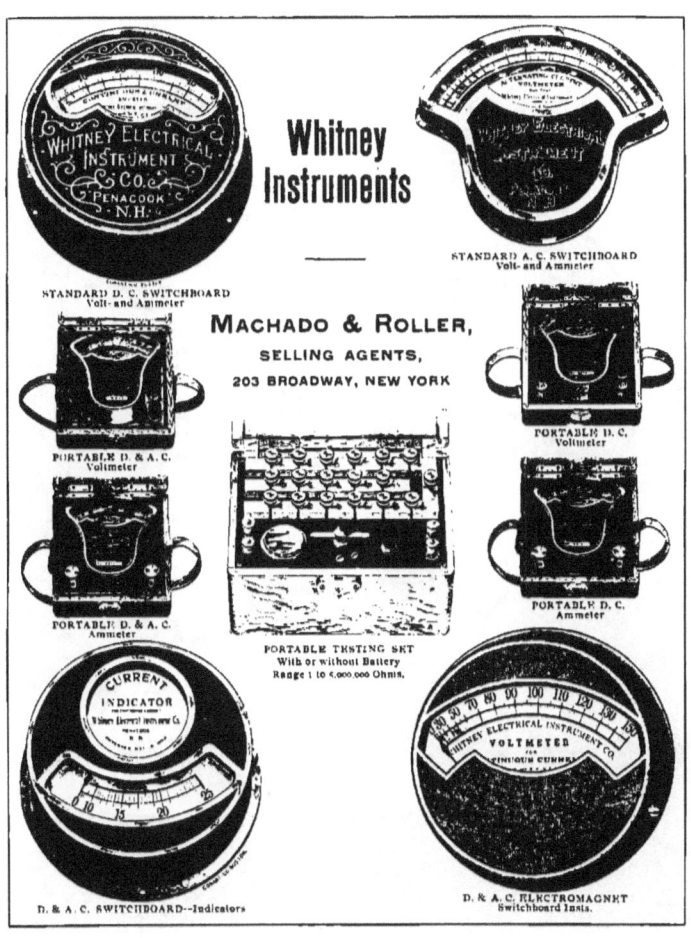

LOUIS E. MASSA
ELECTRICAL ENGINEER

COLLEGE OF PHYSICIANS AND SURGEONS
NEW YORK CITY

INSTALLED BY
FRIEDMAN, RENNARD & CO.

Modern Switchboards

WE MAKE A SPECIALTY OF ALL MODERN
ELECTRICAL APPLIANCES

Electric Lighting
Electric Railway
Electric Power

APPARATUS AND SUPPLIES

We Build and Equip Complete Electric Light
Plants, Electric Railway and Power Plants

BIBBER=WHITE COMPANY

49 Federal Street, Boston

The Eynon=Evans Mfg. Co.

1517=1523 Clearfield St.
Philadelphia, Penna.
..Long Distance Telephone..

MACHINE SHOP AND ENGINEERING ..DEPARTMENT..

We invite correspondence and will cheerfully furnish estimates on Electrical, Steam and Hydraulic Specialties

SEND FOR NEW CATALOGUE OF THE ⟫⟨E⟩⟪ SPECIALTIES

Injectors, Blowers, Exhausters, Ventilators, The "Old Reliable" Steam Trap, Jet Condensers, Syphons and <u>Extra Heavy</u> Globe, Angle and Check Valves 1-4 to 20 in.

Started, Regulated, Stopped with One Handle

Balanced Regulating Valves
Automatic Free Exhaust Valves

The best for Marine, Locomotive and Stationary Boilers

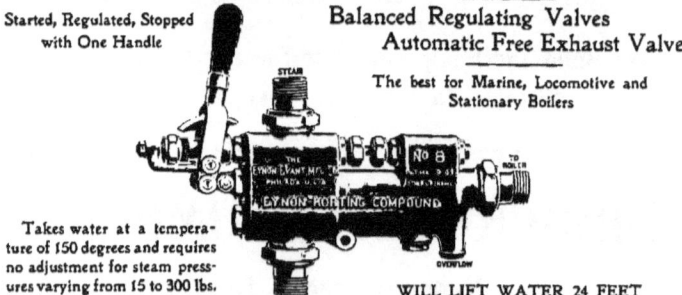

Takes water at a temperature of 150 degrees and requires no adjustment for steam pressures varying from 15 to 300 lbs.

WILL LIFT WATER 24 FEET

EYNON-KORTING COMPOUND INJECTOR

The Eynon=Evans Mfg. Co.

1517=1523 Clearfield St.
Philadelphia, Penna.

..Long Distance Telephone..

COPPER, BRASS AND BRONZE FOUNDRY DEPARTMENT

PURE COPPER CASTINGS

Of highest conductivity, sound and free from blow-holes, easily worked

---

RED and YELLOW BRASS CASTINGS of every description, light or heavy, clean, smooth and accurate to pattern

Switchboard Castings a Specialty

High-Grade Bearing Metals for Engines, Motors, Dynamos, Etc.

PHOSPHOR and MANGANESE BRONZE

PATTERN WORK in all its branches

WRITE FOR ESTIMATES...

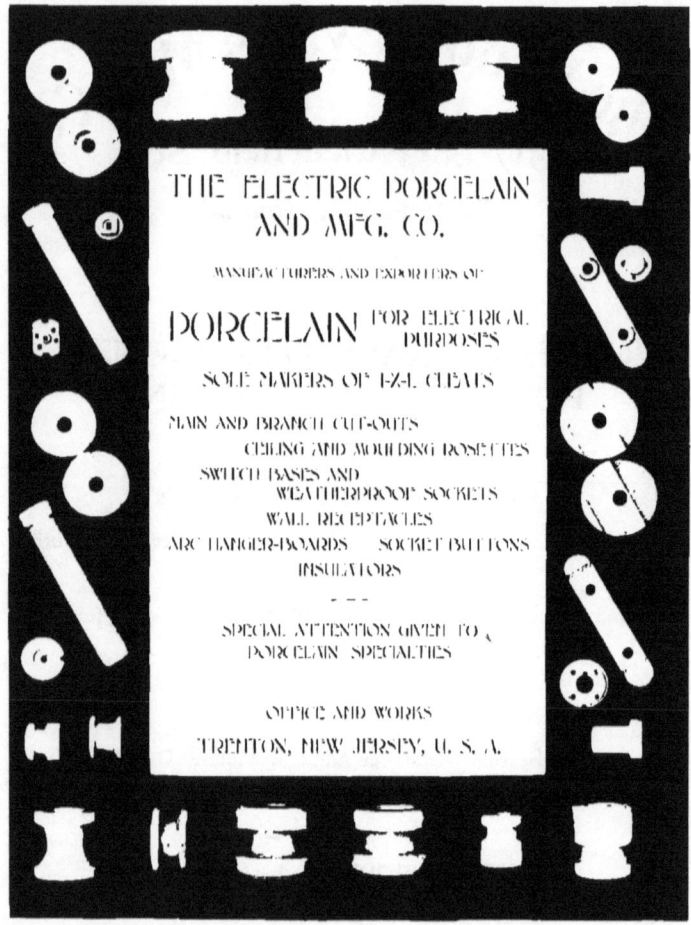

HEIDE BUILDING
NEW YORK CITY

CONTRACTORS AND BUILDERS
McILROY, WARD & CO.

C. J. GOLDMARK
ELECTRICAL ENGINEER

McLeod, Ward & Co.

27 THAMES ST. NEW YORK — Contractors for the Complete Installation of Electric Light and Power Plants, in accordance with the best practice

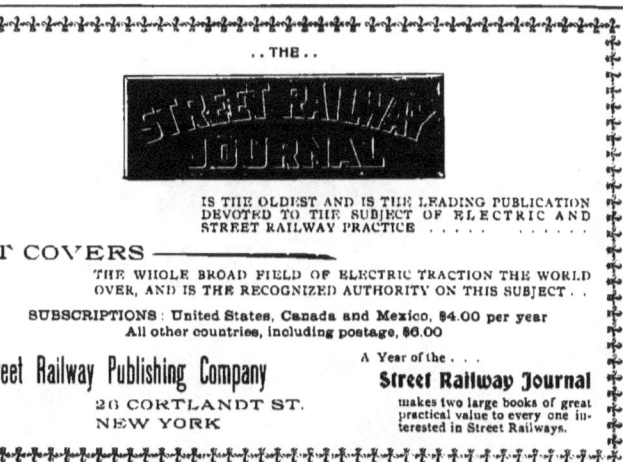

..THE..

STREET RAILWAY JOURNAL

IS THE OLDEST AND IS THE LEADING PUBLICATION DEVOTED TO THE SUBJECT OF ELECTRIC AND STREET RAILWAY PRACTICE

IT COVERS THE WHOLE BROAD FIELD OF ELECTRIC TRACTION THE WORLD OVER, AND IS THE RECOGNIZED AUTHORITY ON THIS SUBJECT..

SUBSCRIPTIONS: United States, Canada and Mexico, $4.00 per year
All other countries, including postage, $6.00

Street Railway Publishing Company
26 CORTLANDT ST.
NEW YORK

A Year of the... **Street Railway Journal** makes two large books of great practical value to every one interested in Street Railways.

HARD RUBBER ELECTRICAL SUPPLIES

MANUFACTURED BY

The Columbia Rubber Works Co.
66 AND 68 READE STREET   **NEW YORK**
FACTORIES AT AKRON, OHIO

"THE DAIKER"
[APARTMENT HOUSE]
NEW YORK CITY

GENL. INC. ARC LIGHT CO.
BUILDERS

N. Y. ELECTRIC EQUIPMENT CO.
CONTRACTORS

The Switches and Switchboards

...OF THE...

General Incandescent Arc Light Co.

NEW YORK

ARE THE RECOGNIZED STANDARDS OF EXCELLENCE AND WORKMANSHIP AND ARE LOW IN PRICE

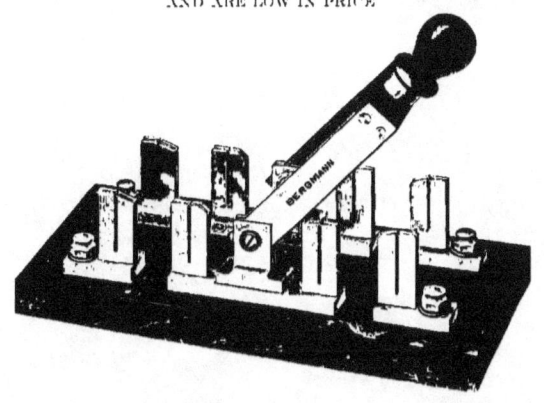

AS EVIDENCED BY ANY OF THE HUNDREDS OF SWITCHBOARDS BUILT BY THEM AND IN USE EVERYWHERE

...Send for Catalogue...

GENERAL INCANDESCENT ARC LIGHT CO.

S. BERGMANN, President

572-578 First Avenue, New York

CORNER THIRTY-THIRD STREET

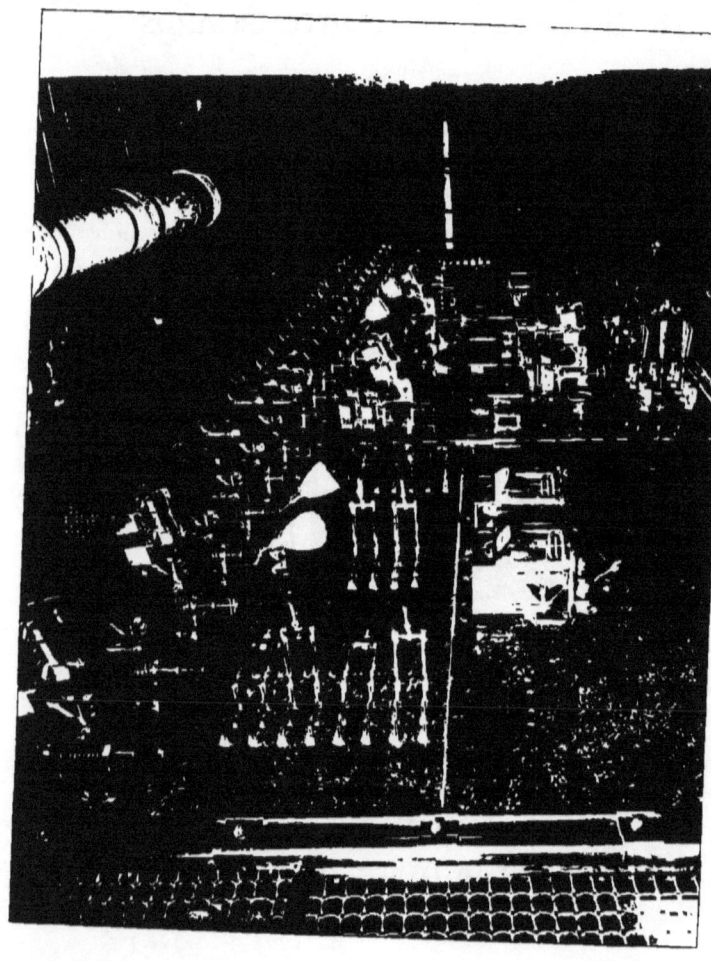

COMMERCIAL CABLE BUILDING
NEW YORK CITY

FRANCIS W. JONES, E. E.
POSTAL TELEGRAPH CO.

CHARLES CUTTRISS, E. E.
COMMERCIAL CABLE CO.

INSTALLED BY
STANLEY & PATTERSON

O. F. ZURN
J. M ZURN
J. D. KELLEY
C. J. CURRAN

THE O. F. ZURN CO.
High-Grade Lubricating Oils and Greases

Particularly suited to Electrical Machinery. If you have Rope Drives,
write for sample of our **ROPOLEUM**, the best Dressing
known for Manilla, Hemp or Cotton Ropes

408-418 VINE STREET   PHILADELPHIA

CARBONS—Electric Light Carbons, Soft Cored and Solid, Carbon Brushes,
Battery Carbons

SOLAR CARBON & MFG. CO.

339 Fifth Avenue, Pittsburgh, Pa.

Manufacturers of EVERYTHING IN THE CARBON LINE

Write for Prices

We are PURCHASERS of the
PLATINUM contained in the base of

BURNED-OUT INCANDESCENT LAMPS

WRITE FOR PARTICULARS **BAKER & CO., Newark, N. J.**

MANUFACTURERS OF PLATINUM SHEET OR WIRE, ANY SIZE, SHAPE
OR DEGREE OF HARDNESS, FOR ALL PURPOSES

"Elephant Brand Phosphor-Bronze"  "Delta Metal"

THE PHOSPHOR BRONZE SMELTING CO. LIMITED.
2200 WASHINGTON AVE., PHILADELPHIA.
"ELEPHANT BRAND PHOSPHOR-BRONZE"
INGOTS, CASTINGS, WIRE, RODS, SHEETS, ETC
— DELTA METAL —
CASTINGS, STAMPINGS AND FORGINGS
ORIGINAL AND SOLE MAKERS IN THE U.S.

CONGRESSIONAL LIBRARY
WASHINGTON, D. C.

INSTALLED BY
UNITED STATES GOVERNMENT
BERNARD R. GREENE, SUPT.

CONSULTING ENGINEER
DR. CAREY T. HUTCHINSON

Built by W. S. Hill Electric Co.

UTICA STATE HOSPITAL, UTICA, N. Y.

W. S. HILL ELECTRIC CO.
NEW BEDFORD, MASS.
BUILDERS AND DESIGNERS OF

Modern Switchboards

Some of our recent installments include

New Public Library, Boston	Willard State Hospital, Willard, N. Y.
Post Office, Boston	Manhattan State Hospital, Ward's Island, N. Y.
City of Boston (4 Boards), Boston	Long Island State Hospital, King's Park, N. Y.
Boston Theatre, Boston	

NOTE.—The Switches used on the Switchboard for the Congressional Library, Washington, D. C., shown on opposite page, are of our manufacture.

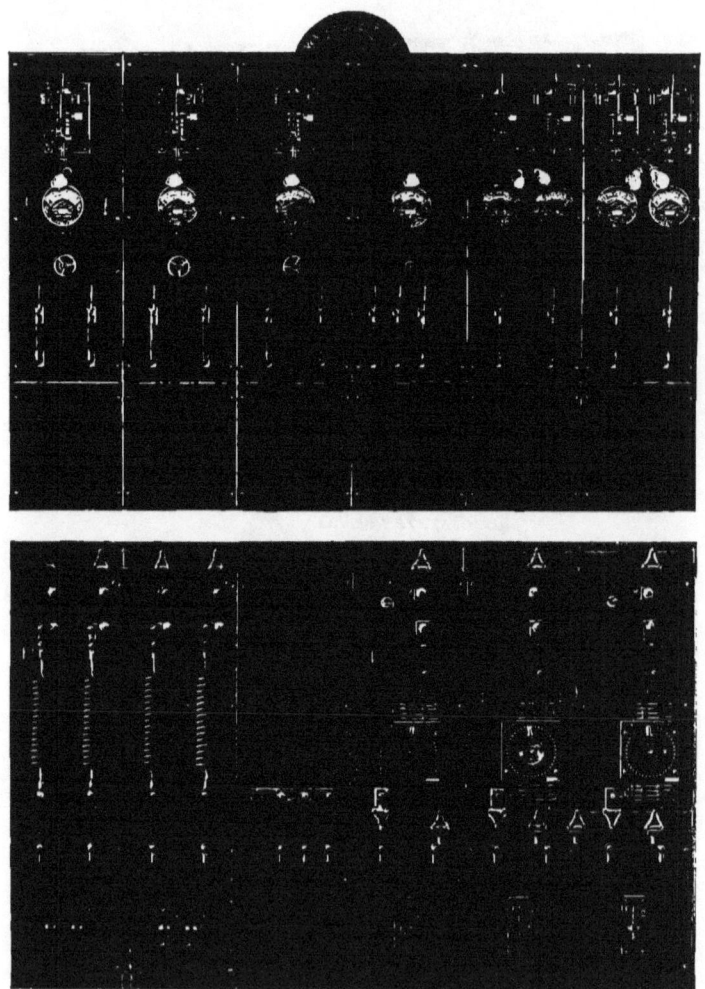

FRONT AND BACK VIEWS OF
SIX-PANEL RAILWAY SWITCHBOARD

BUILT AND INSTALLED BY
WALKER CO.
CLEVELAND, OHIO

Special European Agent for the

Cutter Electrical and Mfg. Co.'s
I-T-E Circuit Breakers

WRITE FOR PRICES AND INFORMATION

ROBERT W. BLACKWELL

39 VICTORIA STREET, W.
London, England

The Crescent Shade

DIAMETER 10 INCHES

This Shade is made of Corrugated Tin, finished in brilliant green enamel outside, and pure white enamel inside. The enamel is baked on and will not flake off or crack.

The Shade is made on dies after our own design, and we have spared no expense to make it the best of its class, and to sell at a price no higher than is asked for inferior goods.

NET TRADE PRICES
Price per dozen, $1.50
Price per gross, $15.00
Special price for larger quantities.

MANUFACTURED BY

VALLEE BROS. & CO.
625 ARCH STREET   **PHILADELPHIA**

AMERICAN LITHOGRAPHIC BUILDING,
NEW YORK CITY

FORD, BACON & DAVIS
ENGINEERS

NEW YORK ELECTRIC EQUIPMENT CO.
CONTRACTORS

New York Electric Equipment Company

S. BERGMANN, President   P. H. KLEIN, Jr., Treasurer

OFFICES AND WORKS
Cor. 33d Street and First Avenue

TELEPHONE
129-38th

MAKE A SPECIALTY OF CARRYING OUT THE SPECIFICATIONS OF ARCHITECTS AND ELECTRICAL ENGINEERS FOR ALL ELECTRICAL WORK, THOROUGHLY AND CORRECTLY, AND WITH A COMPETENT AND COMPLETELY EQUIPPED ESTIMATING DEPARTMENT, FURNISHES ESTIMATES WITH THE GREATEST PROMPTNESS AND ACCURACY

REFERENCES
LEADING ARCHITECTS AND ELECTRICAL ENGINEERS

Agents for

"BERGMANN" LONG LIFE ARC LAMPS
MANUFACTURED BY THE
GENERAL INCANDESCENT ARC LIGHT COMPANY

WEST HALF OF THE LARGEST SWITCHBOARD IN THE WEST
CHICAGO CITY RAILWAY CO.
CHICAGO, ILL.

INSTALLED BY
THE WALKER CO.
CLEVELAND, O.

EAST HALF OF THE LARGEST SWITCHBOARD IN THE WEST
CHICAGO CITY RAILWAY CO.
CHICAGO, ILL.

INSTALLED BY
THE WALKER CO.
CLEVELAND, O.

BOSTON CITY HOSPITAL.

R. B. HATCH
ENGINEER

BUILT AND INSTALLED BY
WM. J. MURDOCK & CO.

The Switchboard shown on the opposite page was made and installed by us at the
BOSTON CITY HOSPITAL

WM. J. MURDOCK & CO.

MANUFACTURERS OF

SWITCHBOARDS

No. 160 CONGRESS STREET

BOSTON, MASS.

We will be pleased to furnish plans and estimates of Switchboards for electrical purposes, and would solicit your correspondence.

WM. J. MURDOCK & CO., 160 Congress Street, Boston, Mass.

LONG DISTANCE TELEPHONE

PATTON BROS.
ELECTRICAL ENGINEERS

BANK OF COMMERCE
NEW YORK CITY

CONTRACTOR AND BUILDER
J. F. HALL

Two Tablets Worth Reading

Get What You Pay For

You advertise to get results.

Results can only be gotten from a paper that reaches the people who buy, or who counsel buying.

The buyers are the managers of central lighting stations, electric railway and power plants, supply houses, isolated plants, telephone exchanges, etc.

These are the people reached by **THE ELECTRICAL ENGINEER.**

QUANTITY of circulation counts only when it means QUALITY too. That is what we have—QUANTITY and QUALITY, and that is why advertising in THE ELECTRICAL ENGINEER pays.

Weekly Circulation, 10,000

Moral

Advertise

Judiciously and Boldly in

THE ELECTRICAL ENGINEER

Get What You Pay For

You read electrical papers to get news.

To get news you want to know the most recent practice in central station design, the latest types of electric generators, motors, dynamos, telephones, and all kinds of electric appliances. You want to know about the great electrical projects of the day.

You want to know the latest electrical supplies on the market, so that whether you may be equipping your own plant, or trying to sell again, you are at least *posted*.

THE ELECTRICAL ENGINEER does this for its readers more thoroughly than any other electrical journal, and that is why it is the most widely read.

10 cents a copy, $3.00 a year.

Weekly Circulation, 10,000

Moral

Read constantly

THE ELECTRICAL ENGINEER

if you desire to

keep abreast of the times.

THE ELECTRICAL ENGINEER
120 LIBERTY STREET, NEW YORK

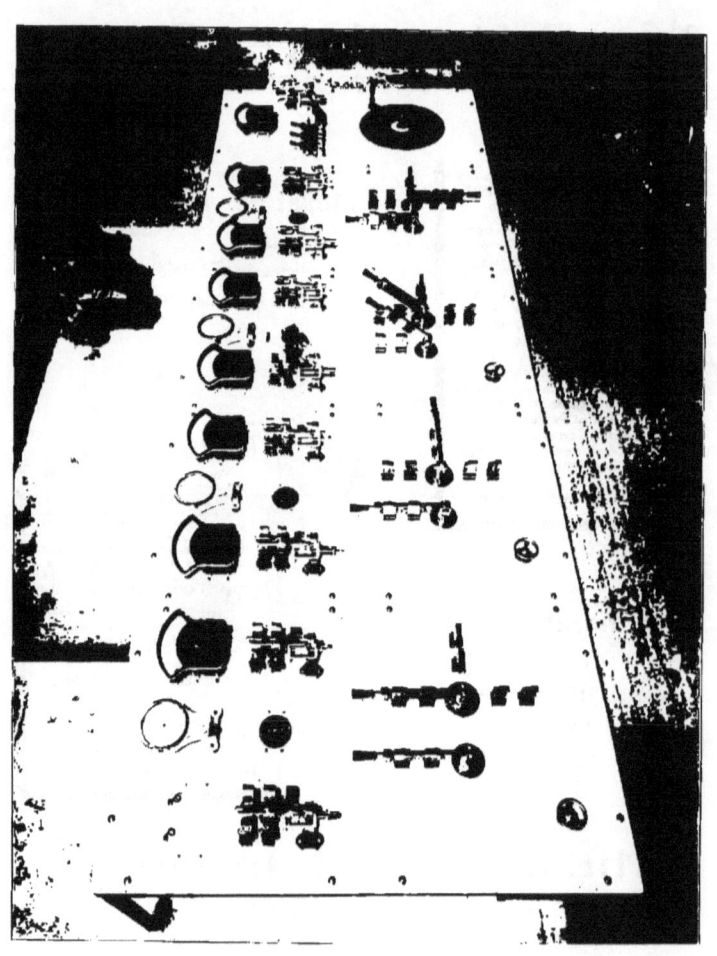

SOME OF THE PUBLICATIONS

...OF...

THE W. J. JOHNSTON COMPANY.

The Electrical World. An Illustrated Weekly Review of Current Progress in Electricity and its Practical Applications. Annual subscription $3.00

Dictionary of Electrical Words, Terms and Phrases. By Edwin J. Houston, Ph.D. Fourth edition. Greatly enlarged. 10,012 words and terms defined; 12.073 definitions; 990 double-column octavo pages; 582 illustrations. An indispensable reference book, not only for electricians, but for every one interested in current progress $7.00

Shop and Road Testing of Dynamos and Motors. By Eugene C. Parham and John C. Shedd. Practical and thorough. 536 pages $2.00

Electro-Dynamic Machinery. By E. J. Houston, Ph.D. and A. E. Kennelly, D.Sc. A text-book on continuous-current dynamo-electric machinery for electric-engineering students of all grades. 431 pages, 232 illustrations $2.50

Practical Calculation of Dynamo-Electric Machines. A Manual for Electrical and Mechanical Engineers, and a Text-book for Students of Electrotechnics. By A. E. Wiener. 683 pages, 375 illustrations $2.50

Gerard's Electricity. With chapters by Dr. Louis Duncan, C. P. Steinmetz, A. E. Kennelly and Dr. Cary T. Hutchinson. Translated under the direction of Dr. Louis Duncan. 392 pages, 112 illustrations. As a beautifully clear treatise for students on the theory of electricity and magnetism, as well as a résumé for engineers of electrical theories that have a practical bearing, the work of Professor Gerard has been without a rival in any language $2.50

Electrical Power Transmission. By Dr. Louis Bell, Ph.D. Uniform with Crosby & Bell's "Electric Railway." Essentially practical in its character. Cloth $2.50

The Theory and Calculation of Alternating Current Phenomena. By Charles Proteus Steinmetz. Contains the very latest knowledge relating to alternate-current phenomena, much of which is original with the author, and here appears for the first time in book form $2.50

Central Station Bookkeeping. With Suggested Forms. By H. A. Foster $2.50

Electric Lighting Specifications for the use of Engineers and Architects. Third edition, entirely re-written. By E. A. Merrill. 213 pages $1.50

Electricity One Hundred Years Ago and To-day. By Edwin J. Houston, Ph.D. 199 pages, illustrated $1.00

Alternating Electric Currents. By E. J. Houston, Ph.D., and A. E. Kennelly, D.Sc. (Electro-Technical Series). . $1.00

Electric Heating. By E. J. Houston, Ph.D., and A. E. Kennelly, D.Sc. (Electro-Technical Series) $1.00

Magnetism. By E. J. Houston, Ph.D., and A. E. Kennelly, D.Sc. (Electro-Technical Series) $1.00

Electro-Therapeutics. By E. J. Houston, Ph.D., and A. E. Kennelly, D.Sc. New edition, enlarged by addition of chapter on X-Rays. (Electro-Technical Series) $1.00

Electric Arc Lighting. By E. J. Houston, Ph.D., and A. E. Kennelly, D.Sc. (Electro-Technical Series) $1.00

Electric Incandescent Lighting. By E. J. Houston, Ph.D., and A. E. Kennelly, D.Sc. (Electro-Technical Series) . $1.00

The Electric Motor. By E. J. Houston, Ph.D., and A. E. Kennelly, D.Sc. (Electro-Technical Series) $1.00

Electric Street Railways. By E. J. Houston, Ph.D., and A. E. Kennelly, D.Sc. (Electro-Technical Series) . . . $1.00

The Electric Telephone. By E. J. Houston, Ph.D., and A. E. Kennelly, D.Sc. (Electro-Technical Series) . . . $1.00

Electric Telegraphy. By E. J. Houston, Ph.D., and A. E. Kennelly, D.Sc. (Electro-Technical Series) $1.00

Alternating Electric Currents. Their Generation, Measurement, Distribution and Application. Authorized American edition. By Gisbert Kapp. 166 pages, 37 illustrations and 2 plates $1.00

Electric Railway Motors. By Nelson W. Perry. 256 pages, many illustrations $1.00

Original Papers on Dynamo Machinery and Allied Subjects. Authorized American Edition. By John Hopkinson, F.R.S. 249 pages, 98 illustrations $1.00

Davis' Standard Tables for Electric Wiremen. With Instructions for Wiremen and Linemen. Rules for Safe Wiring and Useful Formulæ and Data. Fourth edition, Revised by W. D. Weaver $1.00

Experiments with Alternating Currents of High Potential and High Frequency. By Nikola Tesla. 140 pages, 35 illustrations $1.00

Lectures on the Electro-Magnet. Authorized American Edition. By Prof. Silvanus P. Thompson. 287 pages, 78 illustrations $1.00

Dynamo and Motor Building for Amateurs. With Working Drawings. By Lieutenant C. D. Parkhurst $1.00

Reference Book of Tables and Formulæ for Electric Street Railway Engineers. By E. A. Merrill. The practical arrangement of this work, its condensed style and convenient form recommend it to every street-railway engineer . . $1.50

Copies of any of the above books, or of any other electrical book published, will be sent by mail, POSTAGE PREPAID, to any address in the world on receipt of price.

THE W. J. JOHNSTON COMPANY, 253 Broadway, New York.

GERMANIA BUILDING,
MILWAUKEE, WIS.

BUILT AND INSTALLED BY
JULIUS ANDRAE & SONS CO.,
MILWAUKEE, WIS.

A. M. PAITZ
ELECTRICAL ENGINEER

THE OLDEST ELECTRICAL
WEEKLY IN THE
UNITED STATES.......

The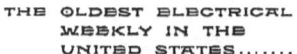

Electrical Review

established 16 years ago, is recognized as the best read and most widely quoted electrical publication in the United States. It is a real NEWSPAPER, published weekly, with illustrations. Its bound volumes form a history of the art.

SUBSCRIPTION RATES:

One Year, United States and Canada,
    post free,   - - - $3.00
One Year, Foreign Countries, - 5.00
Sample Copy, - - - .10

**THE BEST ADVERTISING MEDIUM
IN THE ELECTRICAL FIELD.....**

ELECTRICAL REVIEW PUBLISHING CO.,
Times Building, New York City.

QUEEN'S INSURANCE BUILDING
NEW YORK CITY

C. O. MAILLOUX
ELECTRICAL ENGINEER

J. I. CHAPIN, CONTRACTOR
BUILT AND INSTALLED BY
ZIMDARS & HUNT

SAVE YOUR EYES
BY USING
The Kinsman Desk Lamps and Portables

WE ALSO MANUFACTURE

Van Houten Ceiling Fans, Ward Spark Arresters, Ward Orchestra and Pulpit Lamps, Safety Wire Holders and Modern Switchboards.

Shall be pleased to send descriptive matter on request.

McLEOD, WARD & CO., 27 Thames St., New York

BEWARE OF SPURIOUS IMITATIONS

AMERICAN..
..ELECTRICIAN

AN ILLUSTRATED MONTHLY JOURNAL DEVOTED TO PRACTICAL ELECTRICAL ENGINEERING

THE SCOPE of the American Electrician is broader than that of any other electrical journal in the world. It makes no distinction between the practical man who is a technical graduate, and the practical man who has not had the advantage of technical education.

BOTH need to know the latest developments in the branches in which they are engaged, not only with respect to practice, but also with respect to principles that in time affect practice.

NEITHER cares to have such information burdened with theoretical discussion and mathematical analysis, nor, on the other hand, limited to mere elementary principles.

THE AIM of the American Electrician is to meet these conditions, and that its efforts are appreciated is shown by the fact that it has by far

The Largest Circulation of any Electrical Journal in the Entire World

THE CONTENTS INCLUDE

Authoritative Articles on the very latest electrical developments and advances, both scientific and practical—the treatment being simple without sacrifice of accuracy and instructive to the professed engineer, while coming within the understanding of any intelligent electrical reader.

Practical Departments dealing with central stations, electric railways, interior wiring, steam engineering, telephony, alternating currents, construction of apparatus, electrical measurement, etc. Diagrams of electrical connections, practical hints and kinks, catechism of electricity, queries and answers, etc.

Subscription, $1 per year   The American Electrician Co., Havemeyer Bldg., N. Y.

FAIRCHILD & SUMNER
39-41 CORTLANDT ST. NEW YORK

....GENERAL AGENTS FOR....

ONONDAGA DYNAMO CO., DIRECT CURRENT APPARATUS

THE WARREN-MEDBERY CO.

Manufacturers of the IMPROVED WARREN-ALTERNATING GENERATOR

Imperial Brass Foundry

LIMITED

GEO. H. MOWERY, Manager

BRASS, BRONZE, COPPER
..AND..
WHITE METAL
FOUNDERS

Castings of Red and Yellow Brass, Phosphor-Bronze, White Nickel Metal and Special Metal for Patterns. Fine Castings a Specialty.

Copper Castings of very high conductivity

Special attention given to Electrical Work

No. 119 SPRING STREET
(Formerly Craven Street)

Above Race, between Front and Second Streets

PHILADELPHIA

MERCHANT & CO., Inc.

PHILADELPHIA   NEW YORK   BROOKLYN   CHICAGO

MANUFACTURERS OF AND DEALERS IN

PURE LAKE COPPER
- IN COMMUTATOR BARS,
- ROUND RODS, VARIOUS SIZES
- RECTANGULAR BARS AND ODD SHAPES

SWITCHBOARD SHAPES A SPECIALTY

SHEETS AND PLATES — Brass, Bronze, Copper, Zinc — **WIRE AND RODS**

GERMAN SILVER RESISTANCE WIRE
SPRING BRUSH COPPER

Seamless Drawn Tubing — In Brass, Bronze, Copper

Babbitt and Anti-Friction Metals, Solder, Etc.

PATTISON BROS.   ST. PAUL BUILDING   CONTRACTORS
ELECTRICAL ENGINEERS   NEW YORK CITY   N. Y. ELECTRIC EQUIPMENT CO.

MR. ROBERT W. BLACKWELL

announces that he has opened a branch office in

PARIS, FRANCE

No. 50 Boulevard Haussmann

SIBLEY & PITMAN

TELEPHONE
364 FRANKLIN

59 Duane Street, Corner Elm Street
NEW YORK CITY

Electric Light Supplies

HOUSE WIRING SUPPLIES

AGENTS FOR 　　　　　AGENTS FOR
PARTRICK, CARTER & WILKINS 　 CONNECTICUT TELEPHONE CO.

GENERAL ELECTRIC COMPANY'S SUPPLIES

ALFRED F. MOORE 　　 CHARLES C. KING 　　 ANTOINE BOURNONVILLE

Established 1820

ALFRED F. MOORE

MANUFACTURER OF

Insulated Electric Wire, Flexible Cords and Cables

200 NORTH THIRD STREET
PHILADELPHIA, PA.

HOUSE GOODS
OUR SPECIALTY

1867   1898

Partrick, Carter & Wilkins

Manufacturers and Dealers

125 South Second Street

Philadelphia

GENERAL SUPPLIES
OF ALL KINDS

www.ingramcontent.com/pod-product-compliance
Lightning Source LLC
Chambersburg PA
CBHW031815230426
43669CB00009B/1145